Genetically Modified Organisms - Engineering Life in the 21st Century

Edited by Huseyin Tombuloglu

Published in London, United Kingdom

Genetically Modified Organisms - Engineering Life in the 21st Century
http://dx.doi.org/10.5772/intechopen.1011265
Edited by Huseyin Tombuloglu

Contributors
Bolaji N. Thomas, Charles Oluwaseun Adetunji, Guojun Wang, Ibrahim C. Haznedaroglu, James A. Turner, John R. Caradus, Lijuan Wang, Muyiwa S. Adegbaju, Olalekan Akinbo, Olufunke B. Shittu, Prakriti Karki, Qiang Zheng, Rachel Shibalira, Ruth Asheazi Manzah, Sudarshan GC, Sunday Igu Rock Akile, Xue Wang, Yongkang Xie

First published in London, United Kingdom, 2026 by IntechOpen
IntechOpen is the global imprint of INTECHOPEN LIMITED, registered in England and Wales, registration number: 11086078, 167-169 Great Portland Street, London, W1W 5PF, United Kingdom

For EU product safety concerns: IN TECH d.o.o., Prolaz Marije Krucifikse Kozulić 3, 51000 Rijeka, Croatia, info@intechopen.com or visit our website at intechopen.com.

British Library Cataloguing-in-Publication Data
A catalogue record for this book is available from the British Library

Genetically Modified Organisms - Engineering Life in the 21st Century
Edited by Huseyin Tombuloglu
p. cm.

This title is part of the Genetics Book Series, Volume 15
Topic: Genomics
Series Editor: Kenji Ikehara
Topic Editor: Irina Vlasova-St. Louis

Print ISBN 978-1-80631-283-2
Online ISBN 978-1-80631-282-5
eBook (PDF) ISBN 978-1-80631-284-9
ISSN 3049-7094

If disposing of this product, please recycle the paper responsibly.

IntechOpen Book Series
Genetics
Volume 15

Aims and Scope of the Series

"Genetics," which has been proud of its tradition since Mendel presented his research results in 1865, initially progressed quite slowly due to simple observational approaches of individuals and groups. However, the discovery of double-stranded DNA by Watson and Crick about 70 years ago triggered rapid progress in life sciences, including genetics, which was primarily conducted using Escherichia coli and bacteriophages infecting E. coli. Subsequently, genetics has achieved remarkable developments, such as understanding genetic disorders, including cancers, through research on the biogenesis and differentiation of plants and animals.
The two topics of this book series - Human Genetics, and Genomics - will address important areas of advancement in genetics.
Human Genetics: After fundamental genetics, initially studied with the main goal of revealing the functions of individual genes and proteins, genetics expanded from understanding the genetic system itself to understanding many infectious diseases caused by bacteria and viruses. Consequently, human beings are now overcoming infectious diseases by developing medicinal chemicals, including antibiotics and vaccines. However, genetic disorders remain challenging to cure up to now. Nevertheless, even the cure for them, including various cancers, is coming closer to reality due to the rapid progress of human genetics. In this way, the welfare of human life continues to improve, and even longevity, which was once a dream, has been achieved to some extent in recent years.
Genomics: On the other hand, the understanding of the comprehensive interrelationship of whole genes or whole proteins functioning in one organism has become possible now, as research has entered the era of genomics, owing to the rapid progress of base sequence analysis and bioinformatics. The development of genomics has further made it possible to understand the evolutionary processes of organisms through comparative studies among the genomes of many organisms.
This book series will discuss the findings obtained during the advancement of human genetics and genomics. It is also expected that this series will trigger the formation of a better world composed of human beings and all other organisms on Earth through discussions of research results obtained under the development of general genetics.

Meet the Series Editor

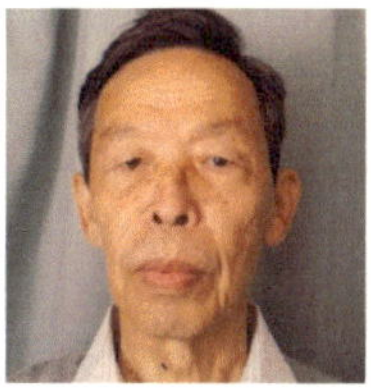

Kenji Ikehara graduated from the Department of Industrial Chemistry, Faculty of Engineering, Kyoto University in 1968. He received his B. Eng. (1968) and subsequently earned M. Eng. (1970) and D. Eng. (1976) degrees from Kyoto University. He began his career as a research associate in the Faculty of Science at the University of Tokyo before moving on to become an associate professor in the Faculty of Science at Nara Women's University. He was later promoted to professor and subsequently served as the dean of the Faculty of Science at Nara Women's University. Additionally, he held the position of director at the Nara Study Center of the Open University of Japan. For approximately 15 years, he focused his research on sporulation initiation of Bacillus subtilis. Later, he shifted his focus to the origins and evolutionary processes of microbial genes, the genetic code, proteins, and life. He has proposed several hypotheses, including the GC-NSF(a) hypothesis on the origin of genes, the GNC-SNS hypothesis on the genetic code, the protein 0th-order structure hypothesis on the origin of proteins, and the [GADV]-protein world hypothesis (GADV hypothesis) on the origin of life. Furthermore, he served as the local chair of the International Conference, Origin 2014, held in Nara in 2014.

Meet the Volume Editor

Huseyin Tombuloglu is an Associate Professor and research scientist at the Institute for Research and Medical Consultations (IRMC) at Imam Abdulrahman bin Faisal University (IAU) in Saudi Arabia. He holds a Ph.D. in Biotechnology. Dr. Tombuloglu specializes in biotechnology, molecular biology, and genomics, with a primary research focus on developing point-of-care molecular diagnostic platforms using CRISPR-Cas systems and nanotechnology. He serves as a principal investigator on major funded research projects supported by the Saudi-NIH (SNIH) and HEVOLUTION foundations and has authored numerous peer-reviewed scientific publications. His interdisciplinary academic work actively bridges advanced genomic workflows with translational biotechnology and biosafety frameworks to address global health challenges.

Contents

Preface

In an era profoundly shaped by rapid scientific breakthroughs, genetic modification stands as a truly transformative force driving modern biotechnology. From the microscopic precision of nucleotide adjustments to the macroscopic scaling of agricultural food security, our ability to edit the code of life is redefining medicine, industrial manufacturing, and ecological stewardship. This edited volume brings together diverse global perspectives to offer a rigorous analysis of genetically modified organisms (GMOs) across biological realms, while simultaneously tackling the profound ethical, legal, and socio-economic questions that accompany these technologies.

The first section of this volume, *Technologies and Applications*, grounds the reader in cutting-edge laboratory methodologies and current scientific discourse. In Chapter 1: *CRISPR-Cas-Based Editing Technologies in Preclinical Models for Disease Correction*, Prakriti Karki and Sudarshan GC provide an essential overview of how precise gene-editing tools are deployed in lab models to pioneer translational therapies for complex human disorders. Shifting the focus to advanced agricultural and medical biomanufacturing, Chapter 2: *Engineering the Chicken: From Transgenesis to Precision Gene Editing for Novel Biomanufacturing* by Lijuan Wang, Yongkang Xie, Xue Wang, Qiang Zheng, and Guojun Wang highlights the evolution of avian platforms from early transgenesis to modern site-specific edits, showing how poultry can serve as efficient bioreactors for vital therapeutic proteins. Concluding the technical section, Chapter 3: *Current Debate Regarding Laboratory and Natural Origin Hypotheses of SARS-CoV-2: A Critically Balanced Review* by Ibrahim C. Haznedaroglu offers a meticulously objective evaluation of the structural and molecular arguments surrounding the emergence of SARS-CoV-2, demonstrating the critical importance of diagnostic oversight and biosafety in the modern virology landscape.

The second section, *Ethical Considerations*, transitions from empirical applications to the crucial societal framework governing biotechnology. In Chapter 4: *Perspective Chapter: Ethical Issues Associated with Genetically Modified Crops, and Can They Be Alleviated?*, John R. Caradus and James A. Turner address the polarizing debates surrounding agricultural GMOs, systematically weighing ecological risks against global food demands and proposing practical pathways to alleviate public and regulatory anxieties. Expanding this bioethical lens to a regional and global scale, Chapter 5: *IPRs, Bioethics, and Equity: Advancing Genome Editing in Africa's Bio-Economy* by Muyiwa Adegbaju, Olalekan Akinbo, Rachel Shibalira, Olufunke B. Shittu, Ruth Asheazi Manzah, Sunday Igu Rock Akile, Charles O. Adetunji, and Bolaji Thomas explores the intricate nexus of intellectual property rights (IPRs), policy frameworks, and socio-economic equity, offering an imperative blueprint for how developing bio-economies can sustainably adopt genome editing tools.

Ultimately, this volume serves as both a showcase of human ingenuity and a reminder of our collective responsibility. By weaving technical innovation with socio-ethical frameworks, it equips scientists, policymakers, and the general readership with the balanced understanding necessary to responsibly participate in ongoing deliberations about the future of genetic engineering.

I would like to express my deepest gratitude to all the contributing authors for sharing their profound expertise, hard work, and commitment to scientific excellence. Special thanks are also extended to the editorial assistants, peer reviewers, and the publishing team at IntechOpen, whose tireless support and dedication ensured the high academic standard and structural integrity of this printed edition.

Huseyin Tombuloglu
Department of Genetics Research,
Institute for Research and Medical Consultations (IRMC),
Imam Abdulrahman bin Faisal University (IAU),
Dammam, Saudi Arabia

Section 1

Technologies and Applications

Chapter 1

CRISPR-Cas-Based Editing Technologies in Preclinical Models for Disease Correction

Prakriti Karki and Sudarshan GC

Abstract

Genome editing technologies for disease correction are advancing, driven by advances in foundational tools such as CRISPR-Cas, protein engineering, and artificial intelligence. This has led to CRISPR-Cas-derived technologies, known as nuclease editors, base editors, and prime editors, which work differently but help correct diseases such as metabolic disorders, neurological diseases, hematologic diseases, retinal disorders, and cancer. Their efficacy and safety have been tested in preclinical models and have shown promising results, leading to several clinical trials of these technologies for the correction of a wide range of diseases. In this chapter, we focus on these approaches, covering the components involved in editing (e.g., endonucleases and guide RNAs), their mechanisms of action, benefits and limitations, the preclinical models they use before clinical trials, and their clinical applications. As a major concern in translating such strategies is the off-target effect, we also focus on several computational, in vitro, and in vivo strategies for detecting off-targets. Likewise, the chapter is mainly focused on DNA editing strategies, but we briefly discuss their use in RNA editing, which also has clinical applications. Even though their application ranges beyond disease correction, for example, in disease modeling, the chapter focuses on preclinical-stage research on disease correction. We conclude with a discussion of the positive impact and technological limitations, including challenges posed by the early stage of development and high costs, which may currently make such technologies unaffordable and inaccessible globally, even if clinically approved.

Keywords: CRISPR-Cas9, nuclease, base editing, prime editing, preclinical models

1. Introduction: Overview of genome editing

Genomic editing is a molecular technology that facilitates precise sequence modification through insertions, deletions, or base substitutions at the genetic level, and it is essential that the ideal genome-editing tool exhibit robust on-target activity, minimal off-target effects, and targeted delivery.

The traditional gene editing platforms included Zinc finger nucleases (ZFNs), the first generation, and Transcription Activator-Like Effector Nucleases (TALEN), the second-generation gene editing technology, and these relied on custom-engineered proteins to recognize DNA strands, making them complex, labour-intensive, and costly. The evolution of CRISPR-Cas systems, the third-generation genome editing technology, has revolutionised the field with its simplicity, flexibility, scalability, adaptability, high efficiency, and low cost of system engineering [1]. Likewise, its potential in disease correction makes it interesting and is the major aim of this chapter. In 2023, the US FDA approved the first cell-based gene therapy, Casgevy, for the treatment of patients with sickle cell disease. The success of gene editing technology makes it a tool that is adapted in preclinical and clinical trials of several diseases where gene editing could be used to solve. However, this technology is not without challenges [2].

Hence, in this chapter, we focus on CRISPR-based gene editing technologies, their benefits and limitations, and their applications in preclinical models, which will ultimately lead to clinical trials and, if successful, to the patient.

1.1 CRISPR-Cas technology

CRISPR (Clustered regularly interspaced short palindromic repeats) was discovered as a gene-editing tool in 2012 by Dr, Jennifer Doudna and Dr. Emmanuelle Charpentier [3]. Since then, this technology has transformed research to a new level, offering unprecedented novel approaches for the treatment of several human diseases, such as blood and metabolic disorders, muscle disorders, and numerous rare inborn diseases, as well as cancer therapy, epigenome editing, regenerative medicine, etc [4].

CRISPR Immune Mechanism: This system evolved from a bacterial adaptive immune system. When bacteria are infected with a virus, the virus captures and incorporates small fragments of viral or plasmid DNA, known as protospacers, into their genome, along with fragments from other invaders they had encountered earlier. These spacers are arranged between repeated palindromic sequences, and this arrangement is what has given CRISPR its name. Such spacers act as a genetic memory of past infections. During subsequent exposure to the same virus, the CRISPR array is transcribed to produce RNA molecules which guide Cas 9 or similar enzymes to the complementary target of viral DNA to cut the DNA apart therefore turning off the virus. Thus, this natural defense mechanism has formed the basis of CRISPR genome editing [3].

CRISPR-based technologies can introduce modifications at the genomic level, producing permanent, heritable alterations (DNA-targeting editors), or modify RNA transcripts without altering genomic DNA, resulting in transient, reversible modifications (RNA-targeting editors). In this chapter, we focus solely on the DNA editors.

1.1.1 Nuclease-mediated genome editing

1.1.1.1 Introduction

This editing tool employs the most conventional Cas9 nuclease to edit the genome and is widely used in preclinical studies.

1.1.1.2 Components

The CRISPR-Cas9 system consists of two key components:

Cas9 endonuclease: It acts as a molecular scissor to cleave DNA, for which it requires a protospacer adjacent motif (PAM sequence), a short sequence immediately downstream of the target site. SpCas9 is the most widely used Cas9 variant, sourced from the bacterium *Streptococcus pyogenes,* which recognizes the PAM sequence (5'-NGG-3, where N is any nucleotide).

Guide RNA (gRNA): A single guide RNA (sgRNA) is the fusion of i) mature CRISPR RNA (crRNA) – has spacer sequence; 18-20 nucleotides long, complementary to the target DNA and ii) a trans-activating crRNA (tracr RNA). It serves as a scaffold to ensure proper orientation with Cas enzyme [5, 6]. Hence, only the modification of the spacer sequence allows programmable targeting of virtually any genomic locus across different organisms if an appropriate PAM sequence is present.

1.1.1.3 Mechanism of action of the CRISPR system

CRISPR-Cas9 editing occurs in three consecutive steps:

Recognition: The gRNA-Cas9 complex scans the genome to identify a PAM sequence. Upon PAM recognition, Cas9 initiates local DNA unwinding, allowing the gRNA to hybridize with the complementary DNA strand and form an R-loop structure.

Cleavage: Cas9 then produces a site-specific double-stranded break (DSB).

Repair: The induced DSB is repaired by one of two endogenous cellular repair pathways: homology-directed repair (HDR) and non-homologous end-joining (NHEJ). HDR requires an exogenous repair template at the DSB site to create precise genome modifications, and this repair template is useful when researchers want to introduce desired sequence edits into the genome. NHEJ repair introduces small, unintended insertion-deletion mutations (indels) that lead to frameshift mutations or premature stop codons and functions independently of a homologous repair template [7]. However, to improve targeted integration in particularly non-dividing cells as well as dividing cells, Homology-independent Targeted Integration (HITI) was developed, thus enabling site-specific DNA insertion via NHEJ mechanisms [8, 9].

1.1.1.4 Benefits versus limitations

In comparison to other conventional gene-targeting technologies, the feature of guide RNA of ease of optimizing for different experimental conditions, has enabled very robust editing and diverse usage. Thus, the simplicity, efficiency, and cost-effectiveness have increased its use in preclinical and clinical settings.

However, CRISPR Cas has its own limitations, including risk of unintended cleavages resulting in chromosomal rearrangements or mutations at untargeted genomic sites, leading to gene function disruption, genomic toxicity, epigenetic alterations, and even carcinogenesis. Likewise, although HDR applications look promising, it has low efficiency in non-diving cells as it primarily occurs in the S/G2 phase of mitotic and meiotic cells, and NHEJ is more active compared to HDR throughout the cell cycle, including divisions and post-mitotic stage [10–12].

Another limitation of the traditional Cas9 is its dependency entirely on the – NGG – PAM motif, which restricts the range of target sites across the genome. This makes the application of the technology feasible for regions with a sparse or absent specific PAM site in the genome. Hence, to address this challenge, strategies for developing additional Cas9 mutants have been implemented to expand the range of target sites. Such mutants are sourced or engineered to feature either of these properties, such as increased target specificity, recognition of multiple PAM sequences, increased editing efficiency, small cas9 fragment size facilitating delivery, or dual functionality for both DNA and RNA cleavage [4] (**Table 1**).

For instance, to expand PAM-recognition abilities and thus the range of target sites, spCas9 variants are engineered by mutations in its PAM-interacting domain, leading to the development of PAM-less or PAM-relaxed Cas9 variants – for instance, VQR, VRER, VRQR, NG, SpG, and SpRY [21]. SpRY remains a significant milestone as it provides greater flexibility to target disease-causing mutations by eliminating the PAM constraint effectively. Thus, expanding the PAM compatibility with engineered Cas variants has also significantly advanced its therapeutic applications (**Table 2**).

1.1.2 Base editing

1.1.2.1 Introduction

It is estimated that nearly 50% of known pathogenic mutants arise from single-nucleotide substitutions, which disrupt protein-coding sequences and underlie human disease [25]. Thus, technology that corrects single-nucleotide variants, particularly in genetic disorders caused by point mutations, holds strong therapeutic importance (**Table 3**).

Base editing is an offshoot of CRISPR genome editing that enables precise single-nucleotide base changes in DNA or RNA to correct faulty mutations without creating double-stranded breaks and without the need for donor DNA templates. This editing technique was initially described by David Liu's group in 2016, marking a significant leap from nuclease-mediated genome editing. It allows efficient four transition mutations, i.e., C→T, G→A, A→G, and T→C; however, it was unable to correct the eight transversion mutations (C→A, C→G, G→C, G→T,

Source	Variant name	Size	Reference	PAM motifs (5'-3')
Streptococcus pyogenes	xCas9	1,368aa approx.	[13]	-NG-, – GAA-, – GAT-
Staphylococcus aureus	SaCas9	1,082aa	[14]	-NNGRRT – (R=A or G)
Streptococcus thermophilus	St1Cas9 St3Cas9	1,122aa 1,393aa	[15] [16]	-NNAGAAW – (W=A or T) - NGGNG-
Neisseria meningitidis	NmCas9	1,109aa	[17]	-NNNNGMTT – (M=A or C)
Campylobacter jejuni	CjCas9	984aa	[18]	-NNNNRYAC – (R=A or G, Y=C or T)
Brevibacillus laterosporus	BlatCas9	1,092aa	[19]	-NNNNCNAA-
Escherichia coli	CasX or Cas12e	9,880aa approx.	[20]	-TTCN-

Table 1.
Different variants of Cas9.

DOI: http://dx.doi.org/10.5772/intechopen.1015771

Disorder	Disease	Model	Primary target organ	Target gene	Editing strategy	Editor variant	Delivery method	Route of administration	Therapeutic outcomes	References
Metabolic	Glycogen storage disorder Ib	Mice	Liver	*G6pc* pR83C	Mutation correction through HDR donor	SaCas9 nuclease	Dual AAV8, *in vivo*	Systemic IV, Temporal vein	18.4 ± 1.5% indels and 0.7 ± 0.06% HDR mediated correction in *G6pc* alleles in liver	[22]
					Double stranded DNA oligonucleotide insertion ("splice-acceptor knock-in" through NHEJ	Cas9 nuclease	LNP, *in vivo*	Systemic; Retro-orbital sinus infusion	Upto ~4% normal hepatic G6Pase α activity was achieved.	[23]
Neurological	Duchenne muscular dystrophy	Humanised mice	Skeletal and cardiac muscle	DMD exon 52	HITI site-specific integration	SaCas9 nuclease	Dual AAV9	Local tibialis anterior injection and Systemic facial vein injection	Low genomic editing efficiency however, high protein restoration of therapeutic benefit	[24]

Table 2.
Application of CRISPR-Cas9 editing in preclinical models.

Disorder	Disease	Model	Primary target organ	Target gene	Editing strategy	Editor variant	Delivery method	Route of administration	Therapeutic outcomes	References
Metabolic	Alpha-1 antitrypsin deficiency	Mice (PiZ transgenic)	Liver (hepatocytes)	*SERPINA1* c.1,096 G > A, (p.E342K)	Point mutation correction	eNme2. C-TadA9e ABE	All-in-one AAV8; in vivo	Systemic (tail vein injection)	22.83% on-target editing in liver at 8 weeks	[26]
Neurological	Duchenne muscular dystrophy	Humanized Mice (DMDΔmE50-51,KIE50/Y)	Skeletal muscle	Dmd (human DMD exon 50 splice regulatory region)	Mutation insertion to splice site to induce exon 50 skipping	NG-ABE8e	Split-intein based dual AAV9; in vivo	Local (Intramuscular injection)	Exon 50 skipping achieved by >90% with 96% dystrophin restoration	[27]
Hematological	Sickle cell dis-ease/ B-thaleassemia	Humanised mice	Bone marrow	HBB codon 6 HbS GTG-> Hb G-Makassar GCG (Ala)	Point mutation from sickle codon to benign variant	ABE8e-NRCH (R=A or G; H=A, C or T)	Helper-dependent adenoviral vector HDAd5/35++	Systemic IV	34% PBMCs with ≥1 corrected allele	[28]
Retinal	Stargardt disease	Humanized mice (Abca4 hu1961E/ ms1961G (KO))	Retina, RPE	*Abca4* c.5,882G > A, (p. G1961E)	Point mutation correction	ABE8.5 m	Split-intein based dual AAV9-PHP.eB; in vivo	Local (subret-inal injection)	A7 gDNA editing efficiency in retina was 31% while A7 mRNA editing was 52%	[29]
Pulmonary	Cystic fibrosis (CF)	Patient-derived rectal organoids	Intestinal	*CFTR* c.2,657 + 5G > A	Splice-site correction	NG-ABEmax mRNA	Electroporation, ex vivo	-	~75% of correct transcripts, controlled bystander edit	[30]

Table 3.
Application of base editing in preclinical disease models.

A→C, A→T, T→A, and T→G) [31]. Such transversion mutations account for approximately 20% of the >75,000 known pathogenic variants [32].

1.1.2.2 Components of CRISPR base editing

Base editors are broadly categorized as cytosine base editors (CBEs) and adenine base editors (ABEs). The components of base editors include:

Modified Cas9 variants in the form of nickase Cas9 (nCas9, D10A) or dead Cas9 (dCas9), which are created by introducing mutations in one or both of the two catalytic domains of Cas9. These lack double-strand cleavage activity. dCas9 facilitates DNA binding without cleaving the backbone and nCas9 causes a single-strand break.

1.1.2.2.1 Deaminase enzyme

Cytidine base editor (CBE) employs cytosine deaminases such as APOBEC1 or its new variants (A3A-D) to convert cytidines (C) to uridines (U), which polymerases interpret as thymines (T). The inherently unstable nature of uracil is normally removed by the cellular repair enzyme uracil DNA glycosylase (UNG). Thus, UNG inhibitor proteins are used in the conventional CBEs to block UNG activity. However, recent advances have shown that the presence of UNG activity with intent can direct new edits. Thus, CBE fused to UNG to create cytosis glycosylase base editors (CGBEs), facilitating not only transition but also transversion, for instance, C to A substitution in bacteria and C to G substitution in mammalian cells [33].

Adenine base editor (ABE) utilises adenosine deaminase, such as TadA, a laboratory-evolved *Escherichia coli* tRNA adenine deaminase, to catalyze the conversion of adenines (A) to inosines and then to guanines (G). The latest deaminase-free glycosylase-based guanine base editor (gGBE) overcomes the limitation of transition-only editing, enabling G-to-C or G-to-T transversions [34].

gRNA: Different generations of base editors have been developed to improve efficiency and minimize indels, such as for cytidine base editor, first generation CBE1, second generation CBE2, third generation CBE3, which exhibited a 6-fold increase in editing efficiency compared to CBE2, and fourth generation CBE4 has been developed. Likewise, different versions of ABEs also progressed through directed and phage-assisted continuous evolution, of which ABE8e is an ABE evolved through protein evolution and engineering that has been shown to increase editing efficiency; this family has become a go-to editor in practical molecular biology and therapeutic applications [35].

1.1.2.3 Mechanism of action

Upon PAM recognition, catalytically impaired Cas9 unwinds the local DNA. It facilitates hybridization of the gRNA with the complementary target strand, while the non-target strand in an R-loop becomes single-stranded and exposed. This exposed ssDNA falls within the defined activity window of the fused deaminase enzyme. Then, the deaminase enzyme catalyzes the deamination of the nucleotide, facilitating its conversion. For instance, cytosine to uracil and adenine to inosine, thus producing U.G and I.T mismatch. Then, the Cas9 nickase introduces a single-strand nick in the non-edited strand, allowing the cellular mismatch and base excision repair pathways to use the edited strand as the template during repair. Thus, during the cellular repair, the cell reads uracil as thymine and inosine as guanine, resulting in permanent C.G to A.T or A.T to C.G substitution [36].

1.1.2.4 Benefits and limitations

Base editing avoids cleavage of the nucleic acid backbone and enables single-nucleotide conversion with high precision and minimal off-targets. Also, it does not require a repair DNA template. However, one major limitation is its inability to catalyze transversion mutation. In addition, base editing happens within the activity window for therapeutic correction, but bystander edits can occur in the presence of additional editable bases within this window. Activity window includes the spacer region with high editing efficiency, and it varies with the type of base editor, which spans a narrow range of bases in the protospacer region where the gRNA binds, for instance, CBEs within a ~5-nucleotide (nt) and ABEs within a ~4-9 nt editing window. Differences in DNA configuration, such as chromatin architecture, depending on locus or cell type, can influence the activity window [37]. Therefore, engineering the base editor partially facilitates narrowing the editing window length [38]. Nonetheless, narrowing comes with the increased requirement of targetable base editors across a broader range of PAM sequences [38].

1.1.3 Prime editing

1.1.3.1 Introduction

Unlike conventional CRISPR-Cas9-mediated gene editing, prime editing enables targeted DNA editing without requiring double-strand breaks or donor DNA templates, as in base editing. It uses a catalytically impaired nCas9 (H840), fused to an engineered reverse transcriptase (RT) and a specialized prime editing guide RNA (pegRNA), which not only directs the editor to the target site but also encodes the desired genetic alteration. The prime editing approach allows precise installation of all transition and transversion mutations, which are 12 in number, in both mitotic and non-mitotic cells, with fewer indels and off-target mutations (**Table 4**). This tool was also pioneered by David Liu and his team three years later, after base editing [31, 39].

1.1.3.2 Components of the prime editing system

Prime editing system is also a two-component system like the above-mentioned editing systems.

1.1.3.2.1 nCas9-Reverse transcriptase fusion complex

Prime editor includes an RNA-protein complex formed by the fusion of RT with programmable Cas9 H840 nickase. RT enzyme transcribes an RNA sequence from

Disorder	Disease	Model	Primary target organ	Target gene	Editing strategy	Editor variant	Delivery method	Route of administration	Outcome	References
Metabolic	Phenylketonria	Humanized liver	Liver (hepatocytes)	*PAH* c.835 T- > C p.F263S	Point mutation correction	PEmax, PE7	Full LNP-mRNA	Tail vein injection	Less than 0.1% editing efficiency with PEmax while PE7 increased the editing rates to 8% in hepatocytes	[40]
Neurodegenerative	Alzheimer disease	Humanized APOE3 mice	Brain	Apoe (p.R136S)	Install the putatively protective coding variant	v3em PE3	Split Npu-intein based dual AAV9; in vivo	ICV injection at P1	DNA editing efficiencies in neocortex 12%, hippocampus 14%	[41]
Hematologic	Sickle cell disease/ B-thalassemia	Humanized mouse	Hematopoietic stem cells	*HBB*	Point mutation correction	PE5max	Helper dependent adenoviral vector (HDAd5/35++); in vivo	Systemic IV	~40%and 43.6% editing by week 6 and 16 post HSC transduction. An average of 43% correction of HbS > HbA.	[42]
Retinal	Retinitis pigmentosa	Mouse	Retina	*PDE6B* exon 13, c.1,678C > T, (p.R560C)	Point mutation correction	PE-SpRY	Split Npu intein mediated dual AAV8	Local (Subretinal injection)	~41% editing in the whole retina with ~0.04% indels and improved visual function.	[43]
Pulmonary	Cystic fibrosis (CF)	Patient-derived rectal organoids	Rectal organoids-intestinal epithelium	*CFTR* c.680 T > G (p.L227R)	Point mutation correction	PE3b	Lentiviral delivery; ex vivo	-	80% functional rescue achieved.	[44]

Table 4.
Application of prime editing in preclinical disease models.

reverse transcription template to a single-stranded DNA sequence. Different variants of prime editor like, PE1 has RT fused to C terminus of Cas 9 nickase through a flexible linker, PE2 has engineered reverse transcriptase, while PE3 is built on the base of PE2 which introduces a second nick in the non-edited DNA strand through a sgRNA [45]. Moreover, PEmax has codon optimization and additional nucleolar localization signals to improve specificity and targeting efficiency (29), and PE7 has the RNA-binding factor La to protect pegRNA from exonuclease degradation [46].

1.1.3.2.2 pegRNA

pegRNA consists of a 3' primer-binding site (PBS), an RT template (RTT) with the edit sequence, sgRNA scaffold that binds to Cas 9, and a 5' spacer (guide) sequence.

1.1.3.3 Mechanism of action

As mentioned earlier, for CRISPR Case 9 and base editing, the process follows the steps of DNA recognition, cleavage (nicking), and repair, but with a few additional steps between nicking and repair.

Upon the recognition of the target genomic site, the spacer sequence of pegRNA hybridizes with DNA and directs the nCas9 nuclease to unwind dsDNA, forming an R-loop, thereby exposing a short segment of ssDNA of the non-complementary strand. Next, the nuclease introduces a nick in the non-target PAM strand, following which the PBS in the pegRNA base pairs with the 3' end and primes the RT to transcribe its template sequence into DNA. This creates a 3' flap that contains the homology region downstream of the nick, allowing it to anneal to the target strand. At the same time, the endogenous cellular DNA repair pathways remove the original 5' flap. However, this heteroduplex branched intermediate DNA, with one edited strand and the other original, unedited strand, forms a mismatched alignment. This is resolved by the mismatch repair (MMR) system, which either preserves or removes the edit. Therefore, the PE3 system was introduced to include an additional sgRNA that nicks the non-edited strand and biases MMR toward the edited strand [45].

1.1.3.4 Benefits and limitations

The two important advantages of this prime editing technique have been its ability to edit DNA sequences without creating double-stranded breaks, thus reducing off-target mutations and frequent indels. In addition, unlike base editing, which creates single-nucleotide changes with few bystander edits, it installs a range of edit sequences in both dividing and non-dividing cells without bystander edits. The major difference from base editing is its necessity to customise and optimize pegRNA and the PE itself [47]. However, studies have reported the low editing efficiency and locus-dependent editing efficiency of prime editors [40]. Likewise, the large size of the editor makes it challenging to deliver the payload and requires development of efficient delivery strategies.

2. RNA editing

The deployment of CRISPR-Cas9 for the manipulation of genomes paved the way for RNA editing technologies. Unlike DNA editing, RNA editing exert reversible changes i.e., allows temporal and tunable control over gene expression as well as reduced off target mutations with improved safety profiles. Thus, it opens avenues for diseases not yet addressed by traditional therapies and for studying gene functions that require precise temporal modulation.

It employs the CRISPR-Cas13 system, which is characterized by the presence of a single RNA-guided RNA endonuclease, Cas13 protein. Cas13-based RNA editing systems have been used for applications such as virus RNA detection, RNA knockdown, RNA base editing, splicing modulation, and transcript labeling [48]. Similar to DNA base editors, catalytically inactive Cas13 is fused with the adenosine or cytidine deaminase domain. In particular, the RNA Editing for Programmable A-to-I (REPAIR) system induces adenosine-to-inosine deamination [49], and the RNA editing for Specific C-to-U Exchange (RESCUE) system, as well as the C to U RNA Editor (CURE), facilitate cytosine-to-uracil deamination [50, 51].

Even though the CRISPR-Cas technology is highly impactful and has challenges, one of the major common challenges is delivering it to cells in vivo and ex vivo to make it work, which is critical. Hence, in the next section, we discuss how such genome editors are delivered to cells in vivo or ex vivo.

3. Delivery of genome editors in vivo

Efficient genome editing requires precise delivery of editing components to targeted cell types, tissues, or organs. Unintended editing in off-target tissues or sites may result in adverse outcomes. Even though delivery strategies may be less limiting in early preclinical systems, such as cell lines or in silico models, successful therapeutic application relies on safe and efficient in vivo delivery. Nevertheless, ex vivo approaches, in which editing of patient-derived stem cells is followed by reinfusion, are equally clinically valuable as delivery would be less of a problem.

Several delivery modalities have been explored and employed to deliver genome editors to target disease cells while preventing their accumulation in healthy or bystander cells. These are broadly categorized as viral and non-viral vectors. The most commonly employed viral vectors are adeno-associated viruses (AAVs) and retroviruses. AAVs are non-integrating vectors that efficiently and transiently express the transgene through long-lived episome formation. Their characteristic persistence, tissue-specific tropism across various AAV serotypes, and low immunogenicity have made them attractive for gene therapy. However, their episomal persistence leads to gradual vector dilution in rapidly dividing cells, thus reducing long-term transgene expression in highly proliferative tissues. Moreover, their applications are constrained by pre-existing neutralizing antibodies [52] and by challenges in large-scale, high-titre AAV production. Also, the limited packaging capacity of ~4.7 kb [53] and the difficulty of large genome editors fitting into a single AAV are additional critical challenges. To overcome this constraint, dual AAV systems that independently express Cas9 proteins and gRNAs have been developed [54]. Moreover, split-intein-based dual AAV delivery has been the commonly employed strategy for delivering larger editors, such as base, prime, and epigenome editors. Herein, the genome editor is split into two fragments: an N-terminal half and a C-terminal half, and reconstituted intracellularly via split-intein-mediated protein splicing, by fusing the N-terminal fragment to the C-intein and the C-terminal fragment to the N-intein, respectively [55]. The most commonly used inteins in split-Cas9 are Npu intein from Nostoc punctiforme DnaE and Rma intein from Rhodothermus marinus DnaB, owing to their high splicing efficiency. In fact, it is important to identify the optimal robust split site within the editor protein without increasing potential off-targets [56]. In addition to the split-intein system, other dual-AAV strategies, including overlapping vectors, trans-splicing system, fragmented genome assembly, etc have also been explored to deliver large genetic payloads [57, 58].

Another viral vector, i.e., lentiviruses, is a subset of retroviruses and is an integrating vector. These can express large genetic payloads of approximately <10kb and efficiently transduce both dividing and non-dividing cells, thus expanding their application ex vivo [59]. Gamma retrovirus carried a risk of insertional mutagenesis, as illustrated by a gamma retroviral clinical trial for X-linked SCID, in which four of the nine patients receiving treatment developed T-cell leukemia due to insertional mutagenesis [60]. Thus, lentiviral vectors have been preferred over gamma retroviruses, thanks to their reduced genotoxicity and safety profile [61]. To improve specificity and tropism, further refinements, such as pseudotyping lentiviral vectors, engineering AAV capsids, and using tissue-specific promoters, are being implemented. Both AAVs and LVs are trusted workhorses for efficiently transducing target cells. However, the risk of immunogenicity and unintended off-target genomic editing is still a major concern. Consequently, there has been an increasing attention toward non-viral delivery systems, particularly lipid nanoparticles (LNPs).

LNPs are non-viral vectors that deliver editors in the form of mRNA, plasmid DNA, or ribonucleoprotein (RNP) complex. Unlike viral vectors such as AAV and retroviruses, which are limited by pre-existing immunity, LNPs are not, thus facilitating repeated dosing [62]. Moreover, it is advantageous where a short-lived expression of editors is necessary to prevent cleavage of an already edited target sequence and to mitigate off-target risks. LNPs have a natural tendency to be taken up by hepatocytes via interactions with apolipoprotein (ApoE) and by hepatic low-density lipoprotein (LDL) receptors [63]. This is helpful for liver-targeted genome editing; however, for extrahepatic targeting, it is necessary to bypass liver accumulation. Thus, several approaches to alter the surface charge, size, and lipid composition, as well as physicochemical characteristics, are optimized to enable tunable targeting and overcome this limitation.

4. Preclinical disease models

The outcomes from in vitro genome editing in primary cells are, although important for mechanistic validation, not necessarily predictive of in vivo efficacy or safety [28]. Likewise, there is discordance between the outcomes obtained from animal research and those from clinical trials, which could be due to biological species differences in pharmacokinetics, metabolism, and immune responses. This emphasizes the translational bottleneck in genome editing technologies. Consequently, the U.S. Food and Drug Administration (FDA) mandates robust data from preclinical models that emulate human disease and physiology for an Investigational New Drug (IND) application before proceeding to human clinical trials. In this section, we will discuss preclinical disease models, particularly animal models, new approaches such as the organ-on-chip model, and in silico models.

4.1 Animal models

Different animal models are chosen based on the research target and consideration of how well the model relates physiologically and pathophysiologically to the target. In particular, for research on genetic disorders, the ideal animal model should display a stable phenotype across generations, harbor the same genetic defect, and mimic key hallmarks and progression of the target (e.g., human) pathology [64]. Other factors, such as lifespan, availability, and size, also influence selection. Most commonly used animal models for the generation of disease and drug testing include:

Mice and rats are cost-effective laboratory animal models and better suited for high-throughput testing than large animals such as pigs, monkeys, dogs, and rabbits [65]. However, limitations persist for some diseases in these animal models. For example, a recent base editing study by Muller et al. 2025 in Stargardt disease obtained genomic correction, yet the model failed to recapitulate the phenotype (51) fully. Likewise, most Alzheimer's disease (AD) mouse models mimic familial AD but not the pure sporadic AD, which accounts for over 95% of total cases in humans [66]. In particular, according to the National Institute of Health (NIH), the major limitation of animal models for clinical translation has been the high failure rate (exceeding 92%) of drugs that look promising in animals but fail to become effective for human treatments [67]. Therefore, it is very important that the tested animal recapitulate the disease phenotype, to be considered relevant from a translational point of view.

4.2 New approach methodologies

Recently, the FDA has given preference to new approach methodologies (NAM) to replace preclinical animal models and to address the challenge of recapitulating disease phenotypes in previous models. NAMs include in vitro human-derived systems such as organs-on-chip, computer-based modeling, and in silico approaches, as well as other systems that facilitate the evaluation of pharmacokinetics, immunogenicity, and toxicity with high relevance [68, 69]. These are not only a good alternative to animal models but also considered ethically sound. However, unlike monoclonal antibodies or small-molecule drugs, no approvals have been found for NAMs in genome editing technologies. For instance, SIGX1094, developed with an AI model and validated in organoid models, received IND approval for diffuse gastric cancer [70].

4.2.1 In vitro human-derived systems

It includes miniature systems designed to mimic the key structure, function, and biological complexity of organs. Hence, such systems are good choices for modeling rare disorders and designing personalized therapy. Both organoids and organ-on-a-chip are miniature systems, but organoids are self-organized 3D cell cultures that mainly mimic tissue architecture and function, whereas organ-on-a-chip, 2D or 3D, incorporates tunable microfluidic flow and forces found in vivo on a microfluidic chip. However, these systems pose challenges, including batch-to-batch variation, differences in protocols for generating in vitro human-derived systems, and difficulties incorporating physiologically relevant features, such as vascular structures and diverse cell types. Hence, it is critical to address these challenges before adopting it as an alternative to animal models. Addressing such challenges might make them a first-choice preclinical model, helping reduce clinical trial failures and accelerating the development of gene editing technologies. So far, the use of such systems has been limited to preclinical experiments in animal models of genome editing technologies to obtain robust data [71, 72].

4.2.2 In silico tools and modeling

With the trend toward applying AI and machine learning across sectors, the genome editing sector can also benefit from such in silico technologies. The increase in data generated by several bench science technologies has shown the potential of in silico tools to predict safety, immunogenicity, and pharmacokinetics, and to support the validation of data from other preclinical models or to replace animal models through data and modeling rapidly and cost-effectively. Such tools include Physiologically-Based Pharmacokinetic (PBPK) Modeling, ML and AI Predictive Models, Bioinformatics, and in silico off-target screening, etc [73, 74].

4.2.3 Other innovative platforms

NAMs also include ex vivo human tissues, cell-based screening, refined in vivo methods, etc. Even though leveraging NAMs is widely adopted for assessing genome editing technologies, replacing animal models completely is not relevant in the present context; however, in the future, as NAMs advance, they could be an alternative to animal models and better recapitulate the disease state [69].

5. Applications

5.1 Disease modeling

CRISPR-Cas is used to understand, model, or correct diseases. The flexibility of knocking-down/, out/in makes them suitable for modeling or understanding the disease. As the chapter's main focus is correcting the disease, here we briefly explain disease modeling and then focus on disease correction. However, disease modeling is an important application of genome editing technology in preclinical trials, as the goal of disease correction in patients cannot be achieved without a preclinical model that mimics human disease. For instance, in one study, prime editing was used to introduce cancer-related mutations into human organoids to model cancer, but with challenges such as off-target effects and undesired mutations at the target site. In the same study, base editing was found to be more relevant for modeling cancer mutations, given its high efficiency and specificity, than prime editing [75]. The therapeutic application of genome editing not only involves modeling and correcting mutations in genetic diseases but also modulating gene expression in the context of both common and complex diseases [76].

5.2 Disease correction

There could be two approaches to correcting disease using genome editors: one is to remove cells from the patient, deliver genome editors into them, and transfer them back to the body, known as ex vivo delivery. Second, by directly delivering the genome editors in vivo. To elaborate further, we take examples of some diseases and discuss how genome editors are used to correct them [77] (**Tables 2–4**).

5.2.1 Metabolic disorder: Glycogen storage disease Ia

Glycogen storage disease-Ia (GSD1a), an autosomal recessive metabolic disorder, is caused by a deficiency of glucose-6-phosphatase-α (G6Pase-α) activity. One of the pathogenic variants prevalent in these diseased patients is the G6PC-R83C variant, resulting in severe fasting hypoglycemia, hepatomegaly, and other metabolic phenotypes. The studies conducted in homozygous knock-in models of G6pc-R83C based on nuclease-based gene editing were performed through two different strategies: one HDR in neonates and the other NHEJ in 2 to 4-week-old mice utilizing AAV [22] and LNP [23], respectively, for the delivery of editors. Both studies demonstrated enhanced survival to adulthood without hypoglycemic seizures. However, HDR-mediated editing was less efficient in non-dividing cells, while NHEJ-mediated editing relied on efficient insertion in the correct position and orientation, and the editing efficiency in both strategies was less than 2%. Then, the same team used a humanised mouse model, huR83C, to mimic the disease and employed BEAM-301, LNP-mediated delivery of gRNA and mRNA of engineered ABE. This approach achieved an editing efficiency of ~60% in the liver, while achieving a therapeutic physiological level of G6Pase-α activity with just ~10% base editing rate [78].

5.2.2 Neurodegenerative disease: Duchenne muscular dystrophy

Duchenne Muscular Dystrophy is a fatal genetic disorder affecting male live births due to genetic mutations in the *DMD* gene of the X chromosome. The idea of micro-dystrophic gene therapy evolved through the concept of Becker muscular dystrophy [79], which aims to express a truncated yet miniaturized partially functional form of dystrophic protein. This approach has facilitated overcoming the limitation of AAV to deliver beyond 4.7kb, since the full-length (FL) DMD gene cDNA exceeds 11kb. However, this approach also reports the failure to provide complete protection to heart and skeletal muscle integrity when translated in clinical studies [80]. One study by Oliver et al. 2021 employed a CRISPR-Cas9-based HITI approach for the restoration of full-length dystrophin in a humanised male mouse model (hDMDΔ52/mdx), which achieved relatively low genome editing efficiency yet high dystrophin protein restoration [24].

5.2.3 Cancer

In cancer, gene editing could be used to directly modify tumor cells or immune cells for efficient cellular immunotherapy. For instance, CRISPR-edited CAR T cells are in preclinical and clinical trials. Nuclease-based editing has been leveraged to transfer the gene to a precise location or to knock out genes that affect its efficiency. Likewise, such technologies have been employed to directly target cancer cells in vivo to restore tumor suppressor genes or disable oncogenes. In several studies, the application of PE and base editing has also been explored for modeling cancer and correcting the mutations. For instance, universal pegRNAs were developed to correct endogenous KRAS mutations in human cancer cells using prime editors and successfully corrected the G13D KRAS mutation to wild-type sequences without indel mutations [75]. Likewise, based editing also corrected oncogenic KRAS and TP53 mutations in patient-derived organoids [81]. These examples clearly demonstrate the application of such techniques in future precision oncology.

5.2.4 Retinal disorder: Retinitis pigmentosa (RP)

Inherited retinal diseases are caused by mutations in genes that encode enzymes, regulate gene expression, or perform structural roles within the eye. Studies are focused on correcting mutations in enzymes that catalyze biochemical reactions for light detection or phototransduction, while others focus on proteins that provide structural support to photoreceptors. Likewise, rhodopsin (Rho) is a light-sensitive protein that provides not only structural but also signaling properties to the cell. A study by Du et al. (2024) employed a base editing strategy in vivo, delivered ABEmax via dual AAV vectors, and yielded up to 44% Rho correction in Rho-E150K homozygous mice [82]. Likewise, prime editor PESpRY has been tested in the diseased retinas of a Pde6b-associated RP mouse model using a dual AAV approach, and has been found to rescue photoreceptors and produce functional PDE6β [43].

5.2.5 Hematologic disease: Sickle cell anemia

Sickle cell anemia is a blood disorder caused by a mutation in a single nucleotide base of the β-globin gene, HBB, from adenine to thymidine at codon 6, resulting in the formation of sickle beta-globin protein in red blood cells. Different approaches have been applied to correct the mutation using CRISPR-based gene therapy. The base editing approach has been successful in alleviating these disease symptoms *in vivo* in mice [83]. However, the inability of this approach to change mutated sickle codon GTG to GAG, the studies have utilized alternative i.e.,to change the thymine nucleotide to a cytosine, thus resulting in a GCG, which is a non-pathogenic variant of HBB, Makassar β-globin (HBBG), and have achieved therapeutically relevant levels of editing with >30% of HbS converted to HbG [28]. In another approach, the same author infused prime editing vectors into the bone marrow of the same mouse model, and, in another study, correction of 43% of hemoglobin S alleles on average and <1% undesired indels was achieved [42]. However, the use of an adenoviral vector in their study was a concern because of its high risk of inflammatory responses.

6. Strategies for detecting off targets

Even though CRISPR-Cas has a wide range of applications in disease correction, its major limitation is off-target effects. Off-target effects are DNA modifications at unintended genomic sites that look alike but are not identical to the target sequence. This leads to chromosomal rearrangements, gene disruptions, oncogene activation, or tumor suppressor silencing. Hence, it is crucial to apply the strategies to identify such unintended effects to ensure their efficacy for disease-correction applications. Among the strategies applied so far, they are categorized as computational, in vitro, and in vivo assays.

Computational strategies leverage algorithmic models to predict off-target genomic sites from CRISPR/Cas editing. It includes tools like Cas-OFFinder [13], CHOPCHOP [14], CRISPOR [15], CasOT [16] and CCTop [17]. Such tools evaluate algorithmic criteria such as how closely the sgRNA sequence aligns with non-target DNA, the binding dynamics of sgRNA-DNA interactions, and the stability of DNA segments near a wide range of applications in disease correction. However, these models are limited by the lack of consideration of biological factors such as epigenetics, chromatin accessibility, and 3D genome structure, underscoring the importance of experimental validation [18].

In vitro off-target detection methods include Digested Genome sequencing (Digenome-seq) [19], Selective Integration of Targeted Endonuclease sequencing (SITE-seq) [20], Circularization for High-throughput Analysis of Nuclease Genome-wide Effects by sequencing (CHANGE-seq) [26], Circularization for In vitro Reporting of Cleavage Effects by sequencing (CIRCLE-seq) [27]. Unlike in silico prediction tools, which screen similar sequences, in vitro detection assays are highly sensitive and unbiased. It leverages purified genomic DNA from living cells. However, these assays also do not account for chromatin accessibility and stability, repair dynamics, epigenetic modifications, etc., as the computational approach does. They produce a higher rate of false-positive off-target detections and require a large number of sequencing reads per sample due to background noise from randomly sheared DNA. In addition, this technique is more labor-intensive and time-consuming [18].

Various methods for in vivo off-target identification have also developed. For example, Breaks Labeling and Enrichment on Streptavidin and Sequencing (BLESS) allows real-time detection of nuclease-induced DSBs in fixed cells by labeling DSBs with biotinylated junctions, capturing the fragments with streptavidin-enriched magnetic beads, and then sequencing them with next-generation sequencing (NGS). It is time sensitive as well as requires high genomic DNA input that corresponds to millions of cells input [29].
The improved versions of BLESS are Breaks Labeling In Situ and Sequencing (BLISS) [30] and immobilized-BLESS (i-BLESS) i-BLESS is an ultra-sensitive method for detection of. which reduces background noise and requires low-input samples in the thousands with high sensitivity [84]. Next, Genome-wide, Unbiased Identification of DSBs enabled by sequencing (GUIDE-seq) is another sensitive and versatile CRISPR/Cas9 off-target assay, well-suited for human gene therapeutic applications. It employs small DNA tags called double-strand oligodeoxynucleotides (dsODN), which are inserted at the cut sites, and off-target effects are evaluated using NGS [41]. Likewise, Discovery of In Situ Cas Off-targets and VERification by sequencing (DISCOVER-seq) employs DNA repair factor, MRE11. It works by detecting MRE11 bound DSBs through

Chromatin Immunoprecipitation Sequencing (ChIP-seq), and later followed by NGS. This method is less sensitive than GUIDE-seq and requires high cell input [44]. Tracking-seq was introduced in 2024 and it has a highly adaptable application for detecting off-target effects from CRISPR/Cas, CBE, ABE, and PE. It works by tracing the binding of replication protein A (RPA) to exposed ssDNA at the cut site, followed by sequencing ssDNA libraries on the specific strand [85].

Numerous tools have been developed to detect off-target events, each offering unique strengths while also carrying its own limitations. Hence, the choice of the tool depends on the cell type, experimental condition, affordability, and the desired sensitivity.

7. Conclusions

In this chapter, we discussed new CRISPR-Cas-based genome-editing technologies, including nuclease editing, base editing, and prime editing, along with their mechanisms and potential applications, primarily for disease correction. Likewise, we discussed preclinical models and delivery strategies used to test such technologies before entering clinical trials. We also touched on new preclinical models, an alternative to animal models that could replace them in the future. With advances in miniaturization and organ-mimicking for testing, advanced experimental and measurement techniques, and AI, genome editing technologies will surely advance and expand.

The chapter discusses the potential of these technologies in treating diverse diseases; however, it is also worth noting that the same technology may not be suitable for all applications. Nevertheless, the ongoing advancement of CRISPR-Cas and the current stage of several of these technologies in clinical trials demonstrate their potential. Given the nascent stage of these technologies, there is still room for improvement, and the question of their accessibility and global impact remains. We cannot ignore the pace at which CRISPR-Cas technology is advancing in genome editing for disease correction.

Author details

Prakriti Karki[1*] and Sudarshan GC[2]

1 Learning Planet Institute, Paris, France

2 The Open University, Buckinghamshire, UK

*Address all correspondence to: prakriti.karki@learningplanetinstitute.org

References

[1] Gupta RM, Musunuru K. Expanding the genetic editing tool kit: ZFNs, TALENs, and CRISPR-Cas9. Journal of Clinical Investigation. 2014;**124**(10):4154–4161. DOI: 10.1172/JCI72992

[2] Singh A, Irfan H, Fatima E, Nazir Z, Verma A, Akilimali A. Revolutionary breakthrough: FDA approves CASGEVY, the first CRISPR/Cas9 gene therapy for sickle cell disease. Annals of Medicine & Surgery. 2024;**86**(8):4555–4559. DOI: 10.1097/MS9.0000000000002146

[3] Jinek M, Chylinski K, Fonfara I, Hauer M, Doudna JA, Charpentier E. A programmable dual-RNA–Guided DNA endonuclease in adaptive bacterial immunity. Science. 2012;**337**(6096):816–821. DOI: 10.1126/science.1225829

[4] Azeez SS, Hamad RS, Hamad BK, Shekha MS, Bergsten P. Advances in CRISPR-Cas technology and its applications: Revolutionising precision medicine. Frontiers in Genome Editing. 2024;**6**:1509924. DOI: 10.3389/fgeed.2024.1509924

[5] Adli M. The CRISPR tool kit for genome editing and beyond. Nature Communications. 2018;**9**(1):1911. DOI: 10.1038/s41467-018-04252-2

[6] Macarrón Palacios A, Korus P, Wilkens BGC, Heshmatpour N, Patnaik SR. Revolutionizing in vivo therapy with CRISPR/Cas genome editing: Breakthroughs, opportunities and challenges. Frontiers in Genome Editing. 2024;**6**:1342193. DOI: 10.3389/fgeed.2024.1342193

[7] Yang Y, Xu J, Ge S, Lai L. CRISPR/Cas: Advances, limitations, and applications for precision cancer research. Frontiers in Medicine. 2021;**8**:649896. DOI: 10.3389/fmed.2021.649896

[8] Maresca M, Lin VG, Guo N, Yang Y. Obligate ligation-gated recombination (ObLiGaRe): Custom-designed nuclease-mediated targeted integration through nonhomologous end joining. Genome Research. 2013;**23**(3):539–546. DOI: 10.1101/gr.145441.112

[9] Cristea S, Freyvert Y, Santiago Y, Holmes MC, Urnov FD, Gregory PD, et al. In vivo cleavage of transgene donors promotes nuclease-mediated targeted integration. Biotechnology and Bioengineering. 2013;**110**(3):871–880. DOI: 10.1002/bit.24733

[10] Chang HHY, Pannunzio NR, Adachi N, Lieber MR. Non-homologous DNA end joining and alternative pathways to double-strand break repair. Nature Reviews Molecular Cell Biology. 2017;**18**(8):495–506. DOI: 10.1038/nrm.2017.48

[11] Liao H, Wu J, VanDusen NJ, Li Y, Zheng Y. CRISPR-Cas9-mediated homology-directed repair for precise gene editing. Molecular Therapy-Nucleic Acids. 2024; **35**(4):102344. DOI: 10.1016/j.omtn.2024.102344

[12] Yeh CD, Richardson CD, Corn JE. Advances in genome editing through control of DNA repair pathways. Nature Cell Biology. 2019;**21**(12):1468–1478. DOI: 10.1038/s41556-019-0425-z

[13] Bae S, Park J, Kim JS. Cas-OFFinder: A fast and versatile algorithm that searches for potential off-target sites of Cas9 RNA-guided endonucleases.

Bioinformatics. 2014;**30**(10):1473–1475. DOI: 10.1093/bioinformatics/btu048

[14] Montague TG, Cruz JM, Gagnon JA, Church GM, Valen E. CHOPCHOP: A CRISPR/Cas9 and TALEN web tool for genome editing. Nucleic Acids Research. 2014;**42**(Web Server issue):W401–407. DOI: 10.3389/fgeed.2024.1342193

[15] Haeussler M, Schönig K, Eckert H, Eschstruth A, Mianné J, Renaud JB, et al. Evaluation of off-target and on-target scoring algorithms and integration into the guide RNA selection tool CRISPOR. Genome Biology. 2016;**17**(1):148. DOI: 10.1186/s13059-016-1012-2

[16] Xiao A, Cheng Z, Kong L, Zhu Z, Lin S, Gao G, et al. CasOT: A genome-wide Cas9/gRNA off-target searching tool. Bioinformatics. 2014;**30**(8):1180–1182. DOI: 10.1093/bioinformatics/btt764

[17] Stemmer M, Thumberger T, Del Sol Keyer M, Wittbrodt J, Mateo JL. CCTop: An intuitive, flexible and reliable CRISPR/Cas9 target prediction tool. PloS One. 2015;**10**(4):e0124633. DOI: 10.1371/journal.pone.0124633

[18] Xu YY, Zhou SM, Wang LY, Zhang R, Li K, Qian ZY, et al. Methods for detecting off-target effects of CRISPR/Cas9. Biotechnology Advances. 2026;**86**:108750. DOI: 10.1016/j.biotechadv.2025.108750

[19] Kim D, Bae S, Park J, Kim E, Kim S, Yu HR, et al. Digenome-seq: Genome-wide profiling of CRISPR-Cas9 off-target effects in human cells. Nature Methods. 2015;**12**(3):237–243. DOI: 10.1038/nmeth.3284

[20] Cameron P, Cameron P, Settle AH, Fuller CK, Thompson MS, Cigan AM, et al. SITE-Seq: A genome-wide method to measure Cas9 cleavage. Protocol Exchange. 2017, DOI: 10.1038/protex.2017.043

[21] Kweon J, Yoon JK, Jang AH, Shin HR, See JE, Jang G, et al. Engineered prime editors with PAM flexibility. Molecular Therapy. 2021;**29**(6):2001–2007. DOI: 10.1016/j.ymthe.2021.02.022

[22] Arnaoutova I, Zhang L, Chen HD, Mansfield BC, Chou JY. Correction of metabolic abnormalities in a mouse model of glycogen storage disease type Ia by CRISPR/Cas9-based gene editing. Molecular Therapy. 2021;**29**(4):1602–1610. DOI: 10.1016/j.ymthe.2020.12.027

[23] Samanta A, George N, Arnaoutova I, Chen H, Mansfield BC, Hart C, et al. CRISPR/Cas9-based double-strand oligonucleotide insertion strategy corrects metabolic abnormalities in murine glycogen storage disease type-Ia. Journal of Inherited Metabolic Disease. 2023;**46**(6):1147–1158. DOI: 10.1002/jimd.12660

[24] Pickar-Oliver A, Gough V, Bohning JD, Liu S, Robinson-Hamm JN, Daniels H, et al. Full-length dystrophin restoration via targeted exon integration by AAV-CRISPR in a humanized mouse model of Duchenne muscular dystrophy. Molecular Therapy. 2021;**29**(11):3243–3257. DOI: 10.1016/j.ymthe.2021.09.003

[25] Porto EM, Komor AC, Slaymaker IM, Yeo GW. Base editing: Advances and therapeutic opportunities. Nature Reviews Drug Discovery. 2020;**19**(12):839–859. DOI: 10.1038/s41573-020-0084-6

[26] Lazzarotto CR, Malinin NL, Li Y, Zhang R, Yang Y, Lee G, et al. CHANGE-seq reveals genetic and epigenetic effects on CRISPR-Cas9

genome-wide activity. Nature Biotechnology. 2020;**38**(11):1317–1327. DOI: 10.1038/s41587-020-0555-7

[27] Tsai SQ, Nguyen NT, Malagon-Lopez J, VV T, MJ A, JK J. CIRCLE-seq: A highly sensitive in vitro screen for genome-wide CRISPR-Cas9 nuclease off-targets. Nature Methods. 2017;**14**(6):607–614. DOI: 10.1038/nmeth.4278

[28] Li C, Georgakopoulou A, Paschoudi K, Anderson AK, Huang L, Gil S, et al. Introducing a hemoglobin G-Makassar variant in HSCs by in vivo base editing treats sickle cell disease in mice. Molecular Therapy. 2024;**32**(12):4353–4371. DOI: 10.1016/j.ymthe.2024.10.018

[29] Crosetto N, Mitra A, Silva MJ, Bienko M, Dojer N, Wang Q, et al. Nucleotide-resolution DNA double-strand break mapping by next-generation sequencing. Nature Methods. 2013;**10**(4):361–365. DOI: 10.1038/nmeth.2408

[30] Yan WX, Mirzazadeh R, Garnerone S, Scott D, Schneider MW, Kallas T, et al. BLISS is a versatile and quantitative method for genome-wide profiling of DNA double-strand breaks. Nature Communications. 2017;**8**(1):15058. DOI: 10.1038/ncomms15058

[31] Anzalone AV, Randolph PB, Davis JR, Sousa AA, Koblan LW, Levy JM, et al. Search-and-replace genome editing without double-strand breaks or donor DNA. Nature. 2019;**576**(7785):149–157. DOI: 10.1038/s41586-019-1711-4

[32] Zhi S, Chen Y, Wu G, Wen J, Wu J, Liu Q, et al. Dual-AAV delivering split prime editor system for in vivo genome editing. Molecular Therapy. 2022;**30**(1):283–294. DOI: 10.1016/j.ymthe.2021.07.011

[33] Zhao D, Li J, Li S, Xin X, Hu M, Price MA, et al. Glycosylase base editors enable C-to-A and C-to-G base changes. Nature Biotechnology. 2021;**39**(1):35–40. DOI: 10.1038/s41587-020-0592-2

[34] Xu F, Zheng C, Xu W, Zhang S, Liu S, Chen X, et al. Breaking genetic shackles: The advance of base editing in genetic disorder treatment. Frontiers in Pharmacology. 2024;**15**:1364135. DOI: 10.3389/fphar.2024.1364135

[35] Cabré-Romans JJ, Cuella-Martin R. CRISPR-dependent base editing as a therapeutic strategy for rare monogenic disorders. Frontiers in Genome Editing. 2025;7:1553590. DOI: 10.3389/fgeed.2025.1553590

[36] Rees HA, Liu DR. Base editing: Precision chemistry on the genome and transcriptome of living cells. Nature Reviews Genetics. 2018;**19**(12):770–788. DOI: 10.1038/s41576-018-0059-1

[37] Gaudelli NM, Komor AC, Rees HA, Packer MS, Badran AH, Bryson DI, et al. Programmable base editing of A•T to G•C in genomic DNA without DNA cleavage. Nature. 2017;**551**(7681):464–471. DOI: 10.1038/nature24644

[38] Kim YB, Komor AC, Levy JM, Packer MS, Zhao KT, Liu DR. Increasing the genome-targeting scope and precision of base editing with engineered Cas9-cytidine deaminase fusions. Nature Biotechnology. 2017;**35**(4):371–376. DOI: 10.1038/nbt.3803

[39] Davis JR, Banskota S, Levy JM, Newby GA, Wang X, Anzalone AV, et al. Efficient prime editing in mouse brain, liver and heart with dual AAVs. Nature Biotechnology. 2024;**42**(2):253–264. DOI: 10.1038/s41587-023-01758-z

[40] Rothgangl T, Tálas A, Ioannidi EI, Weber Y, Böck D, Matsushita M, et al. Treatment of a metabolic liver disease in mice with a transient prime editing approach. Nature Biomedical Engineering. 2025;**9**(10):1705–1718. DOI: 10.1038/s41551-025-01399-4

[41] Tsai SQ, Zheng Z, Nguyen NT, Liebers M, Topkar VV, Thapar V, et al. GUIDE-seq enables genome-wide profiling of off-target cleavage by CRISPR-Cas nucleases. Nature Biotechnology. 2015;**33**(2):187–197. DOI: 10.1038/nbt.3117

[42] Li C, Georgakopoulou A, Newby GA, Chen PJ, Everette KA, Paschoudi K, et al. In vivo HSC prime editing rescues sickle cell disease in a mouse model. Blood. 2023;**141**(17):2085–2099. DOI: 10.1182/blood.2022018252

[43] Qin H, Zhang W, Zhang S, Feng Y, Xu W, Qi J, et al. Vision rescue via unconstrained in vivo prime editing in degenerating neural retinas. Journal of Experimental Medicine. 2023;**220**(5): e20220776. DOI: 10.1084/jem.20220776

[44] Wienert B, Wyman SK, Richardson CD, Yeh CD, Akcakaya P, Porritt MJ, Morlock M, Vu JT, Kazane KR, Watry HL, Judge LM, Conklin BR, Maresca M, Corn JE, et al. Unbiased detection of CRISPR off-targets in vivo using DISCOVER-Seq. Science. 2019;**364**(6437):286–289. DOI: 10.1126/science.aav9023

[45] Chen PJ, Liu DR. Prime editing for precise and highly versatile genome manipulation. Nature Reviews Genetics. 2023;**24**(3):161–177. DOI: 10.1038/s41576-022-00541-1

[46] Yan J, Oyler-Castrillo P, Ravisankar P, Ward CC, Levesque S, Jing Y, et al. Improving prime editing with an endogenous small RNA-binding protein. Nature. 2024;**628**(8008):639–647. DOI: 10.1038/s41586-024-07259-6

[47] Murray JB, Harrison PT, Scholefield J. Prime editing: Therapeutic advances and mechanistic insights. Gene Therapy. 2025;**32**(2):83–92. DOI: 10.1038/s41434-024-00499-1

[48] Zhao L, Qiu M, Li X, Yang J, Li J. CRISPR-Cas13a system: A novel tool for molecular diagnostics. Frontiers in Microbiology. 2022;**13**:1060947. DOI: 10.3389/fmicb.2022.1060947

[49] Cox DBT, Gootenberg JS, Abudayyeh OO, Franklin B, Kellner MJ, Joung J, et al. RNA editing with CRISPR-Cas13. Science. 2017;**358**(6366):1019–1027. DOI: 10.1126/science.aaq0180

[50] Abudayyeh OO, Gootenberg JS, Franklin B, Koob J, Kellner MJ, Ladha A, et al. A cytosine deaminase for programmable single-base RNA editing. Science. 2019;**365**(6451):382–386. DOI: 10.1126/science.aax7063

[51] Huang X, Lv J, Li Y, Mao S, Li Z, Jing Z, et al. Programmable C-to-U RNA editing using the human APOBEC 3A deaminase. The EMBO Journal. 2020;**39**(22):e104741. DOI: 10.15252/embj.2020104741

[52] Manno CS, Pierce GF, Arruda VR, Glader B, Ragni M, Rasko JJE, et al. Successful transduction of liver in hemophilia by AAV-Factor IX and limitations imposed by the host immune response. Nature Medicine. 2006;**12**(3):342–347. DOI: 10.1038/nm1358

[53] Dong JY, Fan PD, Frizzell RA. Quantitative analysis of the packaging capacity of recombinant

adeno-associated virus. Human Gene Therapy. 1996;**7**(17):2101–2112. DOI: 10.1089/hum.1996.7.17-2101

[54] Yang Y, Wang L, Bell P, McMenamin D, He Z, White J, et al. A dual AAV system enables the Cas9-mediated correction of a metabolic liver disease in newborn mice. Nature Biotechnology. 2016;**34**(3):334–338. DOI: 10.1038/nbt.3469

[55] She K, Liu Y, Zhao Q, Jin X, Yang Y, Su J, et al. Dual-AAV split prime editor corrects the mutation and phenotype in mice with inherited retinal degeneration. Signal Transduction and Targeted Therapy. 2023;**8**(1):57. DOI: 10.1038/s41392-022-01234-1

[56] Wei R, Yu Z, Ding L, Lu Z, Yao K, Zhang H, et al. Improved split prime editors enable efficient in vivo genome editing. Cell Reports. 2025;**44**(1):115144. DOI: 10.1016/j.celrep.2024.115144

[57] Sondergaard PC, Griffin DA, Pozsgai ER, Johnson RW, Grose WE, Heller KN, et al. AAV. Dysferlin overlap vectors restore function in dysferlinopathy animal models. Annals of Clinical and Translational Neurology. 2015;**2**(3):256–270. DOI: 10.1002/acn3.172

[58] Halbert CL, Allen JM, Miller AD. Efficient mouse airway transduction following recombination between AAV vectors carrying parts of a larger gene. Nature Biotechnology. 2002;**20**(7):697–701. DOI: 10.1038/nbt0702-697

[59] Dong W, Kantor B. Lentiviral vectors for delivery of gene-editing systems based on CRISPR/Cas: Current state and perspectives. Viruses. 2021;**13**(7):1288. DOI: 10.3390/v13071288

[60] Hacein-Bey-Abina S, Garrigue A, Wang GP, Soulier J, Lim A, Morillon E, et al. Insertional oncogenesis in 4 patients after retrovirus-mediated gene therapy of SCID-X1. Journal of Clinical Investigation. 2008;**118**(9):3132–3142. DOI: 10.1172/JCI35700

[61] Beard BC, Dickerson D, Beebe K, Gooch C, Fletcher J, Okbinoglu T, et al. Comparison of HIV-derived lentiviral and MLV-based gammaretroviral vector integration sites in primate repopulating cells. Molecular Therapy. 2007;**15**(7):1356–1365. DOI: 10.1038/sj.mt.6300159

[62] Kazemian P, Yu SY, Thomson SB, Birkenshaw A, Leavitt BR, Ross CJD. Lipid-Nanoparticle-Based Delivery of CRISPR/Cas9 Genome-Editing Components. Molecular Pharmaceutics. 2022;**19**(6):1669–1686. DOI: 10.1021/acs.molpharmaceut.1c00916

[63] Francia V, Schiffelers RM, Cullis PR, Witzigmann D. The biomolecular corona of lipid nanoparticles for gene therapy. Bioconjugate Chemistry. 2020;**31**(9):2046–2059. DOI: 10.1021/acs.bioconjchem.0c00366

[64] Soufizadeh P, Mansouri V, Ahmadbeigi N. A review of animal models utilized in preclinical studies of approved gene therapy products: Trends and insights. Laboratory Animal Research. 2024;**40**(1):17. DOI: 10.1186/s42826-024-00195-6

[65] Lee H, Yoon DE, Kim K. Genome editing methods in animal models. Animal Cells and Systems. 2020;**24**(1):8–16. DOI: 10.1080/19768354.2020.1726462

[66] Kitazawa M, Medeiros R, M. Laferla F. Transgenic mouse models of alzheimer disease: Developing a better model as a tool for therapeutic

interventions. Current Pharmaceutical Design. 2012;**18**(8):1131–1147. DOI: 10.2174/138161212799315786

[67] Marshall LJ, Bailey J, Cassotta M, Herrmann K, Pistollato F. Poor translatability of biomedical research using animals — A narrative review. Alternatives to Laboratory Animals. 2023;**51**(2):102–135. DOI: 10.1177/02611929231157756

[68] FDA. Pushes to replace animal testing. Nature Biotechnology. 2025;**43**(5):655–655. DOI: 10.1038/s41587-025-02690-0

[69] U.S. Food and Drug Administration (FDA). Roadmap to Reducing Animal Testing in Preclinical Safety Studies [Internet]. Report FDA Doc. No.186092. 2025 April. Available from: https://www.fda.gov/media/186092/download [Accessed: 2026-May-01]

[70] Zhang H, Zhuo S, Zhang L, Li L, Peng W, Jiang F, et al. Discovery of SIGX1094R, a novel FAK inhibitor simultaneously blocking FAK and SRC signaling. Cancer Research. 2025;**85**(8_Supplement_1):3018–3018. DOI: 10.1158/1538-7445.AM2025-3018

[71] Li M, Izpisua Belmonte JC. Organoids — Preclinical Models of Human Disease. New England Journal of Medicine. 2019;**380**(6):569–579. DOI: 10.1056/NEJMra1806175

[72] Ma C, Peng Y, Li H, Chen W. Organ-on-a-Chip: A new paradigm for drug development. Trends in Pharmacological Sciences. 2021;**42**(2):119–133. DOI: 10.1016/j.tips.2020.11.009

[73] Gangwal A, Lavecchia A. Artificial intelligence in preclinical research: Enhancing digital twins and organ-on-chip to reduce animal testing. Drug Discovery Today. 2025; **30**(5):104360. DOI: 10.1016/j.drudis.2025.104360

[74] Mansouri K, Karmaus AL, Fitzpatrick J, Patlewicz G, Pradeep P, Alberga D, et al. CATMoS: Collaborative Acute Toxicity Modeling Suite. Environmental Health Perspectives. 2021;**129**(4):047013. DOI: 10.1289/EHP8495

[75] Geurts MH, De Poel E, Pleguezuelos-Manzano C, Oka R, Carrillo L, Andersson-Rolf A, et al. Evaluating CRISPR-based prime editing for cancer modeling and CFTR repair in organoids. Life Science Alliance. 2021; **4**(10):e202000940. DOI: 10.26508/lsa.202000940

[76] Pickar-Oliver A, Gersbach CA. The next generation of CRISPR-Cas technologies and applications. Nature Reviews Molecular Cell Biology. 2019;**20**(8):490–507. DOI: 10.1038/s41580-019-0131-5

[77] Laurent M, Geoffroy M, Pavani G, Guiraud S. CRISPR-based gene therapies: From preclinical to clinical treatments. Cells. 2024;**13**(10):800. DOI: 10.3390/cells13100800

[78] Arnaoutova I, Aratyn-Schaus Y, Zhang L, Packer MS, Chen HD, Lee C, et al. Base-editing corrects metabolic abnormalities in a humanized mouse model for glycogen storage disease type-Ia. Nature Communications. 2024;**15**(1):9729. DOI: 10.1038/s41467-024-54108-1

[79] England SB, Nicholson LVB, Johnson MA, Forrest SM, Love DR, Zubrzycka-Gaarn EE, et al. Very mild muscular dystrophy associated with the

deletion of 46% of dystrophin. Nature. 1990;**343**(6254):180–182. DOI: 10.1038/343180a0

[80] Hoy SM. Delandistrogene Moxeparvovec: First Approval. Drugs. 2023;**83**(14):1323–1329. DOI: 10.1007/s40265-023-01929-x

[81] Sayed S, Sidorova OA, Hennig A, Augsburg M, Cortés Vesga CP, Abohawya M, et al. Efficient correction of oncogenic KRAS and TP53 mMutations through CRISPR base editing. Cancer Research. 2022;**82**(17):3002–3015. DOI: 10.1158/0008-5472.CAN-21-2519

[82] Du SW, Newby GA, Salom D, Gao F, Menezes CR, Suh S, et al. In vivo photoreceptor base editing ameliorates rhodopsin-E150K autosomal-recessive retinitis pigmentosa in mice. Proceedings of the National Academy of Sciences. 2024;**121**(48):e2416827121. DOI: 10.1073/pnas.2416827121

[83] Newby GA, Yen JS, Woodard KJ, Mayuranathan T, Lazzarotto CR, Li Y, et al. Base editing of haematopoietic stem cells rescues sickle cell disease in mice. Nature. 2021;**595**(7866):295–302. DOI: 10.1038/s41586-021-03609-w

[84] Biernacka A, Zhu Y, Skrzypczak M, Forey R, Pardo B, Grzelak M, et al. i-BLESS is an ultra-sensitive method for detection of DNA double-strand breaks. Communications Biology. 2018;**1**(1):181. DOI: 10.1038/s42003-018-0165-9

[85] Zhu M, Xu R, Yuan J, Wang J, Ren X, Cong T, et al. Tracking-seq reveals the heterogeneity of off-target effects in CRISPR-Cas9-mediated genome editing. Nature Biotechnology. 2025;**43**(5):799–810. DOI: 10.1038/s41587-024-02307-y

Chapter 2

Engineering the Chicken: From Transgenesis to Precision Gene Editing for Novel Biomanufacturing

Lijuan Wang, Yongkang Xie, Xue Wang, Qiang Zheng and Guojun Wang

Abstract

The development of recombinant therapeutic proteins has revolutionized modern medicine, yet traditional mammalian cell-based production systems are often associated with high costs and limited scalability, while nonmammalian systems frequently lack proper protein processing capabilities. This chapter explores the chicken – specifically, the hen oviduct – as a powerful transgenic and gene-editing platform for biopharmaceutical production. The biological rationale leverages the oviduct's evolution as a natural protein factory, capable of producing complex glycoproteins that deposit directly into eggs – sterile, self-contained bioreactors with significant economic and scalability advantages over conventional systems. We trace the evolution of methodologies, from initial transgenesis techniques to clustered regularly interspaced short palindromic repeats (CRISPR)-based precision gene editing via PGCs, which allows for the targeted integration of human protein genes. The core focus is on engineering hens to secrete complex therapeutic proteins, such as monoclonal antibodies or enzymes, directly into their eggs. Case studies demonstrate successful production of functional human proteins, with discussion of purification processes and germline transmission strategies for establishing stable lines. Furthermore, the chapter addresses the critical ethical and biosafety considerations inherent in generating genetically modified animals, including animal welfare, environmental containment, and public perception, alongside regulatory pathways from major authorities such as the Food and Drug Administration (FDA) and European Medicines Agency (EMA). By outlining the current achievements and future trajectory of this technology, including integration with synthetic biology and artificial intelligence (AI)-driven design, this chapter positions the genetically modified chicken not only as a pivotal tool in biomanufacturing but also as a compelling case study in the responsible application of genetic engineering for human health.

Keywords: chicken oviduct bioreactor, precision gene editing, transgenesis, recombinant proteins, CRISPR

1. Introduction: Challenges in current biopharmaceutical production platforms

In the past decades, the number of protein-based drugs approved by the US Food and Drug Administration (FDA) has significantly increased. According to the THPdb2 (updated version of the database of approved therapeutic peptides/proteins, https://webs.iiitd.edu.in/raghava/thpdb2/) database released in 2024, a total of 894 protein and peptide-based drugs have been approved in humans by the FDA to date, including 354 monoclonal antibodies and 85 therapeutic peptides. These biologics are widely employed in the treatment of cancer, infectious diseases, and immune-related disorders, and their rapid expansion has substantially driven the pharmaceutical industry's growing demand for large-scale production of consistently therapeutic-relevant proteins [1].

1.1 Limitations of existing mammalian cell platforms

Biopharmaceuticals refer to recombinant proteins, encompassing monoclonal antibodies as well as nucleic acid and genetically modified cell-based products [2]. Based on the most recent survey of biopharmaceutical approvals (January 2018–June 2022), mammalian cell platforms remain the predominant expression system in commercial use. Among these, Chinese hamster ovary (CHO) cells continue to be the dominant mammalian host, accounting for approximately 89% of individual products made in mammalian expression systems. In addition to CHO cells, several other mammalian cell lines have been used for protein production. These include NS0 murine myeloma cells, baby hamster kidney (BHK) cells, human embryonic kidney (HEK) cells, Sp2/0 mouse myeloma cells, and PER.C6 human retinal cells [2]. However, each of these has only contributed to a limited number of commercially approved products.

Over 20 years ago, the idea of utilizing transgenic animals for synthesizing therapeutic proteins was proposed. Significant advancements have since been made in developing techniques for creating transgenic animals, particularly in expressing therapeutic proteins in the mammary glands of ruminants (cows, sheep, and goats) [3]. Mammalian expression systems, represented by CHO cells, have become the standard platform for producing complex mammalian proteins in biopharmaceuticals. The mammalian expression system requires extremely high preproduction costs for platform construction and operation, and the costs of cell culture media and equipment during the production process are also very high. Furthermore, scaling up production proves to be challenging, often requiring the expansion of production space and additional equipment (**Table 1**). While CHO cells allow basic posttranslational modifications, such as human-like glycosylation, they remain inadequate for certain complex modifications, such as specific types of glycosylation and lipidation [4]. These limitations may lead to a loss of biological activity and pharmacological effects. Despite these drawbacks, mammalian cell-based expression systems remain dominant due to the unwillingness to bear the risks associated with transitioning to alternative platforms [5].

Feature	Mammalian cell platforms	Nonmammalian expression systems	Chicken bioreactors
Production cost	High	Low	Low
Scalability	Limited	High	High
Glycosylation pattern	Human-like	Divergent (e.g., high-mannose in yeast)	Human-like
Posttranslational modifications	Complete	Incomplete or aberrant	Complete
Production timeline	Long	Short	Short

Table 1.
Comparison of different biopharmaceutical production platforms.

1.2 Limitations of nonmammalian expression systems in protein processing and modification

Recent recombinant protein products approved by the FDA and EU, particularly biosimilars, demonstrate that nonmammalian products also account for a significant portion of the market. The most commonly used nonmammalian system is *Escherichia coli* (which accounts for approximately 20% of approved products), followed by *Pichia pastoris*, *Saccharomyces cerevisiae*, and *Pseudomonas fluorescens* [2]. The advantages and disadvantages of *Escherichia coli* are equally notable. It offers lower costs and shorter production cycles, making it highly beneficial for large-scale manufacturing (**Table 1**). However, due to its limited posttranslational modification capabilities, it is mainly suited for the production of relatively simple proteins and cannot meet the demands for producing complex therapeutic glycoproteins [2].

2. Biological rationale: Why chicken bioreactors?

2.1 The hen oviduct as a natural protein secretion factory

Although CHO cells remain the dominant platform for recombinant protein production, efforts to generate recombinant proteins through transgenic animals have never ceased. Over the past two decades, several recombinant proteins derived from transgenic animals have been approved for clinical use by the FDA and EU (**Table 2**). These products have been obtained from transgenic goats, rabbits, and chickens [2, 6]. The chicken oviduct bioreactor is widely recognized as a leading strategy for generating recombinant proteins for use in biomedical research and clinical applications [7].

2.2 Eggs as natural, sterile, and highly efficient bioreactors with glycosylation advantages

Chickens possess several inherent advantages as bioreactors for recombinant protein production. First, eggs represent a natural, safe, and sterile source of proteins. The eggshell and its multilayered membranes provide an effective physical barrier that protects recombinant therapeutic proteins expressed within the egg,

Therapeutic protein	Animal/system	Approval year	Indication	Regulatory status
ATryn (rh antithrombin)	Transgenic goat (milk)	2006 (EU) 2009 (US)	Hereditary antithrombin deficiency	Approved
Ruconest (C1 esterase inhibitor)	Transgenic rabbit (milk)	2010 (EU) 2014 (US)	Acute angioedema	Approved
Kanuma (sebelipase alfa)	Transgenic chicken (egg white)	2015 (EU) 2015 (US)	Lysosomal acid lipase deficiency	Approved
SBC-103 (recombinant a-N-acetylglucosaminidase)	Transgenic chicken (egg white)	–	Mucopolysaccharidoses IIIB	Phase 1/2

Table 2.
Representative therapeutic proteins produced in genetically engineered animals.

thereby reducing the risk of exogenous contamination and potentially enhancing protein stability and *in vivo* half-life. Compared with *Escherichia coli* and CHO cell expression systems, egg-derived recombinant proteins are less susceptible to external contamination during downstream purification processes, contributing to improved biosafety.

Protein glycosylation has emerged as a key regulator in the pathogenesis, progression, and therapeutic response of numerous diseases, such as autoimmunity, cancer, and inflammatory diseases [8]. The characteristics of the IgG glycome are shaped by environmental factors, genetic background, and the body's inflammatory status. Distinct glycosylation patterns can then finely modulate IgG effector functions, thereby contributing to the development and progression of autoimmune and inflammation-associated diseases [8]. As an essential posttranslational modification, glycosylation has also been recognized as a key regulatory factor in type 2 (T2) inflammation and related allergic diseases [9]. Moreover, in oncology, aberrant expression and altered glycosylation of MUC1 can activate multiple signaling pathways, promoting tumor cell migration, invasion, and accelerated tumor growth [10].

At present, transgenic animals are considered the most reliable platforms for producing recombinant glycoproteins with human-compatible glycosylation patterns, whereas prokaryotic expression systems, such as *E. coli*, lack the machinery required for glycosylation. Notably, the glycosylation profiles mediated by the chicken expression system differ from those generated in conventional CHO cell systems, which may confer distinct advantages in certain clinical applications [11].

2.3 Economic and scalability considerations

Eggs have long been recognized as being among the most economical providers of protein available for human nutrition. Ovalbumin, the major protein component used as a carrier in chicken bioreactor systems, accounts for approximately 54% of egg white, which substantially simplifies downstream processing and significantly reduces the cost and complexity of recombinant protein separation and purification. Once a stable transgenic chicken line has been established, the costs associated with animal husbandry are considerably lower than those required for CHO platform construction and cell culture maintenance. The incubation period for chickens is approximately 21 days, and sexual maturity and the onset of egg laying occur at around 6 months of age, allowing the establishment of transgenic poultry production flocks within a relatively short time frame. In contrast, the generation and rearing of lactating transgenic mammals require substantially longer periods, typically about 18 months for goats and up to 3 years for cows before they can be utilized for recombinant protein production.

Moreover, production scale can be rapidly expanded by increasing the breeding population within a relatively short period of time, enabling higher yields at a lower cost and with less technical complexity compared to conventional CHO-based expression systems. In addition, eggs have a long-established history as a critical substrate for human vaccine production, and the well-developed manufacturing standards and regulatory frameworks provide valuable experience and reference for the advancement of therapeutic recombinant protein production.

3. Precision genome engineering enables chicken bioreactors

3.1 Early transgenic strategies and their limitations

In mammalian transgenic research, exogenous gene introduction or genome modification can be achieved through techniques such as pronuclear microinjection into zygotes, somatic cell nuclear transfer (SCNT), and embryonic stem cell (ESC)-mediated gene transfer [12]. These approaches were also explored at an early stage in chicken species. In 1988, Perry et al. successfully cultured isolated single-cell fertilized chicken eggs in vitro until embryonic development was completed [13]. Subsequently, Love et al. [14] introduced target gene DNA into fertilized chicken embryos, and among the hatched offspring, one rooster exhibited germline chimerism, with 3.4% of its progeny inheriting the exogenous gene. Salter et al. [15] injected recombinant chicken leukosis virus into the yolk of unincubated fertilized eggs, achieving stable integration of viral DNA into the germline and generating transgenic chickens carrying retroviral sequences, with germline chimerism efficiencies ranging from 1% to 4%. However, both pronuclear injection and nuclear transfer require the recovery of fertilized eggs through invasive procedures involving euthanasia and abdominal dissection of laying hens, which not only conflict with animal welfare principles but also substantially increase experimental complexity and cost.

Viral vector–mediated gene transfer and embryonic electroporation represent alternative strategies for chicken genetic engineering that utilize early stage embryos outside the maternal body. One key benefit of these approaches is that they allow for the convenient obtaining of large numbers of early embryos without harming the hen. For viral vector approaches, the critical step is the precise delivery of vectors carrying exogenous genes into the blastoderm cavity of unincubated fertilized eggs. Kodama et al. [16] successfully generated transgenic chickens expressing human erythropoietin (hEPO), with high-level expression detected in egg white at high concentrations (1,000–6,000 IU/mL), contrasting with relatively low levels in blood and no detectable hEPO in egg yolk. Similarly, Lee et al. [17] employed lentiviral vectors to achieve expression of the human urokinase-type plasminogen activator gene in both the blood and eggs of gene-edited chickens.

Embryonic electroporation involves the application of carefully controlled pulsed electric fields to early chicken embryos, inducing the formation of nanoscale transient pores in the cell membrane that allow efficient and direct cellular delivery of exogenous DNA. Using this technique, researchers have successfully introduced foreign genes into key embryonic tissues, including neural tissue, limb mesenchyme, and somites, thereby enabling the generation of gene-edited chickens [18]. Véron et al. [19] constructed a clustered regularly interspaced short palindromic repeats (CRISPR)/Cas9 dual-guide RNA system to precisely induce segmental deletion of the PAX7 gene and successfully delivered the system into chicken embryos via electroporation. Although these approaches have been widely applied in early studies of chicken genetic manipulation and have resulted in the production of transgenic chickens, they are generally associated with low transgenic efficiency. This limitation is largely attributable to the oviparous reproductive biology of chickens, in which fertilization and early embryonic development occur within the oviduct.

Consequently, by the time fertilized eggs are laid, embryos have already progressed to the multicellular stage, making true zygotes difficult to obtain. Manipulation at later embryonic stages often misses the optimal developmental windows, such as the morula or blastocyst stages, resulting in low germline transmission efficiency. Although viral vector–based methods substantially improve gene integration efficiency, they are inherently limited by random genomic integration, making it difficult to generate stably inherited, site-specific knock-in transgenic chickens [20]. As a result, many transgenic technologies widely used in mammals are not directly applicable to poultry. The oviparous nature of chickens significantly increases the complexity of embryonic manipulation, causing the development of chicken transgenic technologies to lag far behind those established in mammalian systems.

3.2 CRISPR-/Cas-based genome editing in chicken

Chickens are the first livestock and poultry species to have their whole genome sequenced; however, compared with mammals, the complex structure of fertilized eggs and the distinct organizational forms of chicken embryos make the generation of transgenic chickens relatively challenging [20, 21]. With the widespread application of gene-editing tools like CRISPR-Cas9 and the overcoming of barriers in in vitro chicken breeding techniques, chicken genetic breeding has entered a new phase of development [6, 22].

The CRISPR/Cas system, which stands for Clustered Regularly Interspaced Short Palindromic Repeats and CRISPR-associated proteins, is a powerful genome-editing tool [22]. This system, in cooperation with a programmable sequence of approximately 20 bp, enables targeted and site-specific cleavage of target DNA. The CRISPR-Cas represents a form of adaptive immunity, in which this memory is acquired from exogenous genomes and inserted into the CRISPR array as spacer sequences, namely sgRNA. Gene-editing nucleases, including Cas9, can induce targeted double-strand breaks (DSBs), thereby activating multiple endogenous DNA damage repair pathways [23]. By exploiting different endogenous repair mechanisms according to specific research objectives, distinct gene-editing strategies can be achieved. Currently, gene-editing strategies applied in chickens or chicken cells include gene knockout, gene knock-in, gene replacement, and single-base editing, among others [24]. Previous studies have employed CRISPR/Cas9-mediated DSBs to successfully knock out the chicken ovalbumin (OVA) and ovomucoid (OVM) genes, thereby reducing these two major egg allergens [25]. In addition, large-fragment deletion of the myostatin gene can be achieved, leading to increased meat production in poultry [26]. Recently, researchers have also used CRISPR technology to knock out IFNAR1 and IL1R1, providing a new strategy to enhance chicken immunity and antiviral responses [27].

3.3 Applications of primordial germ cells in chicken

In contrast to mammals, where they form later, chicken primordial germ cells (PGCs), the progenitors of sperm and oocytes, are specified during the very early stages of embryogenesis [28]. In the chicken embryo, PGCs are initially located in the area pellucida of the preincubation blastodisc before migrating anteriorly to colonize the germinal crescent [29]. As embryonic development proceeds, the intraembryonic and extraembryonic vascular systems are established, allowing PGCs to enter these vascular networks and gradually migrate to the upper regions of the embryo, ultimately accumulating in large numbers at the genital ridges located at the posterior end of the embryo. This orderly process of migration and aggregation enables PGCs to be isolated either from the circulation at 2.5 days of embryonic development or from the gonads at 5.5 days, and subsequently cultured in vitro [30].

In 2006, researchers successfully isolated and cultured chicken PGCs in vitro and injected enhanced green fluorescent protein (eGFP)-expressing transgenic PGCs into White Leghorn chicken embryos at H&H stages 13–15 during gastrulation, resulting in germline-chimeric transgenic chickens [31]. Following the injection of transfected PGCs into the bloodstream of recipient chicken embryos at 2.5 days of development, they circulate with the blood flow and eventually colonize the gonads, leading to the production of G0 germline-chimeric chickens. Therefore, PGCs represent an ideal gene-editing vehicle for chicken species.

Although PGC-mediated gene editing exhibits clear advantages in achieving targeted germline chimerism, this approach still faces several challenges, one of which is competition between endogenous and exogenous PGCs. If interference from endogenous PGCs could be effectively eliminated, the proportion of G0 chimeras would be expected to increase significantly, thereby further improving the efficiency of this gene-editing strategy. In addition, whether embryos or PGCs are used, the generation of G0 chimeras is unavoidable. Given that chickens require approximately six months to reach sexual maturity, a large number of experimental animals must be raised to adulthood for screening, which inevitably prolongs the breeding cycle and increases maintenance costs.

Researchers have also explored sperm-mediated gene editing and adult PGC transplantation as alternative approaches to directly obtain edited sperm and shorten the breeding cycle. Harel-Markowitz et al. [32] successfully introduced linearized p-eGFP gene sequences into sperm using liposome-mediated transfection, thereby generating transgenic offspring. Cooper et al. [33] systematically optimized and improved sperm-mediated gene-editing methods with the aim of establishing a more efficient and rapid workflow for producing gene-edited chickens. However, both domestic and international studies on the use of sperm as gene carriers for generating transgenic animals have reported inconsistent experimental outcomes, with relatively low reproducibility, and the underlying mechanisms have yet to be fully and clearly elucidated.

Another alternative strategy involves the generation of sterile recipient roosters for PGC transplantation. Physical (irradiation) or chemical (busulfan) treatments can be applied to the testes of sexually mature roosters aged seven to nine months to ablate endogenous germline stem cells, thereby preventing sperm production. Genetically modified PGCs are then introduced into the testes of the recipient, where they can develop into mature sperm. Several research groups have successfully sterilized recipient roosters using gamma irradiation and subsequently obtained offspring [34–36]. However, due to growing global concerns regarding the supply and safety of radioactive sources, particularly in the context of radiological terrorism, X-ray irradiation has increasingly been adopted as an alternative to gamma rays. Whether X-rays can serve as an equivalent dosimetric substitute for gamma irradiation remains to be further investigated.

3.4 Precise gene editing

Gene knock-in refers to the integration of a donor sequence into a specific genomic locus, thereby enabling the insertion of exogenous sequences or the replacement of endogenous genomic sequences with donor DNA. It is an effective strategy for achieving stable gene overexpression. Early gene knock-in approaches mainly relied on random integration or transposon-based insertion into the host genome. Such methods are highly stochastic and may easily disrupt the function of other genes. In addition, the copy number and insertion sites of the integrated sequences are variable, resulting in pronounced heterogeneity in gene expression among cells derived from the same lineage, which is unfavorable for precise control of gene expression levels.

The CRISPR/Cas9 system allows site-specific integration of exogenous genes, thereby simplifying the genetic background and improving the precision and efficiency of experimental manipulation. Through targeted DSB induction, CRISPR/Cas9 activates cellular DNA damage repair pathways, making complex or fine-tuned targeted genome engineering feasible [37, 38]. Following CRISPR/Cas9-induced DSBs, gene knock-in typically requires the provision of exogenous donor DNA or the use of specific endogenous genomic sequences as repair templates. Donor integration can be achieved via homology-directed repair (HDR) mediated by homologous recombination (HR) or microhomology-mediated end joining (MMEJ) [39, 40]. In cases where homologous repair templates are unavailable, donor sequences can also be integrated into the genome through the relatively efficient nonhomologous end joining (NHEJ) pathway [41]. These three repair pathways have been widely applied for locus-specific insertion across various biological systems, including human cells, murine cells, and monkey embryos [42].

A gene knock-in strategy derived from the NHEJ pathway is termed homology-independent targeted integration (HITI), which has proven effective for introducing exogenous genes in both proliferating and postmitotic cells [43]. However, studies in chicken fibroblast DF-1 cells have revealed that HITI achieves a mere 0.97% integration efficiency [44]. Gene knock-in approaches based on HDR-mediated repair are also referred to as homology-mediated end joining (HMEJ). In 2018, Isao Oishi et al. [45] successfully achieved knock-in of the human interferon beta gene (hIFN-β) at the OVA locus using HDR, resulting in hens that produced eggs containing human interferon. In this study, the donor vector contained a 2,800 bp 5′ homology flank and a 3,200 bp 3′ homology flank [45]. Because HR is highly sensitive to mismatches within homology regions, the probability of HDR-mediated recombination is relatively low.

The MMEJ functions as a major DSB repair mechanism in the early M and S phases, precisely when HDR is inactive [37, 46]. The MMEJ-mediated repair requires only minimal homologous sequences (5–40 bp) for DSB repair and site-specific genome insertion, making donor vector construction considerably simpler [47]. A comparative study demonstrated that in primary chicken embryonic fibroblasts, the knock-in efficiency mediated by MMEJ was 4.7%, whereas that mediated by HDR was only 1.3% [46]. In 2021, Kennosuke Ichikawa et al. [48] employed MMEJ-mediated homology repair to achieve site-specific knock-in of RIG-I into the GTP-binding protein gene Mx in DF-1. Unlike model organisms, poultry genomes exhibit relatively high heterozygosity and a high frequency of single-nucleotide polymorphisms (SNPs), making it difficult to identify long homology arms of several hundred or even thousands of base pairs. Taken together, MMEJ represents a simple and efficient gene knock-in strategy that is well-suited for genome editing in chickens.

4. Translational progress: From engineered eggs to therapeutic proteins

4.1 Egg white as a biomanufacturing output

From a biomanufacturing perspective, eggs should be redefined as highly independent, scalable, and regulatory-friendly biomanufacturing units. Unlike cell culture systems that rely on complex process controls, egg production depends on the highly programmed and reproducible physiological processes of the hen, which naturally ensure stability and predictability. Each egg is an independent, well-contained production event, with the target protein expressed and modified *in vivo* and then noninvasively collected in the form of egg white. This discreteness significantly reduces the risk of cross-batch contamination and simplifies quality traceability.

Meanwhile, this platform provides a scaling-up pathway centered on scalability, enhancing production capacity by increasing the number of egg-laying hens, with costs and risks significantly lower than traditional scale-up models in reactors. Moreover, the composition and volume of egg white from a single egg remain relatively stable, making it easier to establish standardized quality control. More importantly, as a historical medium for vaccine production, eggs have accumulated mature regulatory experience and standards, laying a solid regulatory foundation for their use in therapeutic protein production. Therefore, egg white, with its discreteness, scalability, standardization potential, and regulatory friendliness, becomes the key point connecting transgenic chicken technology with industrial application.

4.2 Case studies of therapeutic proteins produced in eggs

As natural macromolecular bioactive substances, proteins demonstrate immense clinical application prospects in the treatment of various diseases, including tumors, immune-related disorders, and metabolic diseases. Currently, protein-based drugs primarily encompass peptide therapeutics, monoclonal antibodies, recombinant antibodies, and genetically engineered vaccines [30]. The chicken oviduct has been successfully utilized as a bioreactor to express various exogenous proteins, such as human β-interferon, epidermal growth factor, human lysozyme, and erythropoietin

[45, 49–51]. Herron et al. [3] precisely cloned the 0.5 kb coding sequence for IFNα2a into an equine infectious anemia virus(EIAV)-based lentiviral vector that is replication-defective. To further enhance the expression efficiency of IFNα2a, they successfully injected this optimized lentiviral vector into chicken embryos, leading to the generation of transgenic chicken lines.

On December 8, 2015, the FDA approved a novel drug named Kanuma as the first therapy for patients with a rare disease called recombinant human lysosomal acid lipase (rhLAL) deficiency [52]. The production technology for Kanuma is based on a transgenic chicken platform, which generates the therapeutic protein by expressing exogenous rhLAL in chickens. The core of this technology involves ligating the LAL gene fragment into a retroviral vector. The packaged virus is injected into chicken embryos, which, after incubation, yield G0 chimeras, from which G1 transgenic chickens are subsequently obtained. This ultimately enables the specific biosynthesis of rhLAL in the hen oviduct, providing a sustained source of the drug for treatment.

The process primarily involves the following key steps (U.S. Avian Glycosylation Patent Application #20140298503): (1) First, the pALV-SIN-GalT1 vector containing the exogenous gene and the ovalbumin (OM) promoter is constructed to ensure specific LAL expression in the chicken oviduct. This vector is packaged into viral particles and subsequently administered to stage X embryos for gene transfection. (2) Embryos injected with the transgenic vector hatch into G0 chicks after approximately 21 days. Preliminary screening via blood tests determines transgenic chimerism efficiency. Upon reaching adulthood, founders with germline chimerism are identified through DNA testing of rooster semen. These are then crossed with wild-type hens to obtain G1 transgenic offspring. (3) A particularly ingenious aspect of the Kanuma design involved using a similar method to create transgenic chickens expressing chicken α-2,3-sialyltransferase type 3. Crossing homozygous G2 offspring from these two transgenic lines resulted in chickens carrying both exogenous transgenes. This design was implemented to enable the final protein to acquire specific glycosylation modifications, thereby enhancing its clinical efficacy.

The chicken oviduct bioreactor provides a promising, scalable alternative system for manufacturing therapeutic proteins. However, efficient purification of target proteins from eggs remains insufficiently explored and presents significant challenges due to the abundant endogenous proteins in egg whites that contaminate the purification process. Protein A chromatography, combined with ammonium sulfate precipitation and cation-exchange chromatography, removed 84.6% and 93.8% of egg white proteins with 86% and 76% antibody recovery [53]. Although Protein A chromatography can purify monoclonal antibodies, egg white contamination persists, indicating the need to develop scalable purification methods to accommodate future industrial production [53].

4.3 Germline transmission strategies

Over the past 30 years, different research teams have primarily focused on two aspects in their efforts to obtain genetically modified chickens: One is how exogenous genes can be stably integrated into the genome, and the other is how to achieve higher germline transmission efficiency. In the field of mammals, advanced technologies such as the cryopreservation of male sperm, female oocytes, or embryos have effectively maintained the diversity and richness of genetic resources. However, for chickens, the collection and preservation of embryos and oocytes face many complex and severe challenges. Woodcock et al. [54] successfully transplanted the original germ cells of rare genetic chicken breeds into chicken embryos treated with DDX4 gene knockout using embryo vascular injection technology. These specially treated embryos successfully grew to sexual maturity (G0 generation) during subsequent incubation and feeding processes. More importantly, their offspring (G1 generation) completely inherited the unique genetic traits of the rare genetic chicken breeds. This research not only deeply reveals the core role of PGCs in the transmission of genetic information but also effectively validates the feasibility of cryopreserving rare breed chicken PGCs and preserving genetic resources through the innovative application of infertile chicken line technology. This has significant implications for promoting the sustainable development of poultry genetic resources.

In the production process of genetically modified chickens, the ability to culture PGCs marks a crucial milestone achievement. Van de Lavoir et al. [2] pioneered long-term chicken PGCs culture in vitro, employing buffalo rat liver (BRL) or STO cells as the feeder layer. This system successfully maintained the continuous culture of PGCs extracted from embryonic blood in a specific medium for more than 35 days [31]. Crucially, these PGCs, after long-term in vitro culture, still exhibited the ability to reach the gonads and give rise to mature, functional gametes when transplanted into the blood system of recipient embryos [55]. The establishment of PGCs isolation and long-term in vitro culture has proven pivotal, enabling significant advances in the field of genetically edited chicken research and development.

Building on the long-term PGCs in vitro culture system, Schusser et al. [56] used HR technology to edit the immunoglobulin light chain or heavy chain of PGCs and then transplanted these PGCs into the blood system of embryos at the HH13–15 stage, successfully producing chimeras. This was the first successful creation of a nonmammalian vertebrate through the combination of long-term PGC in vitro culture and HR technology to knock out specific genes. Notably, they successfully cultured PGCs in vitro for up to 217 days. Li Bichun et al. [57] reprogrammed chicken embryo fibroblasts into induced iPSCs via the introduction of four factors: Oct4, Sox2, Nanog, and Lin28A. They then induced these pluripotent stem cells to further differentiate into PGCs using BMP4, BMP8b, and EGF, providing a new option for germline transmission in chickens.

4.4 Regulatory readiness and commercial reality

The approval and clinical application of Kanuma have set a great precedent for the development of chicken oviduct bioreactors. Sebelipase alfa (Kanuma) has been proven to be an effective treatment for lysosomal acid lipase deficiency (LAL-D), leading to significant improvements in both biochemical markers and clinical outcomes. Clinical trials have shown that treatment with sebelipase alfa significantly reduces liver transaminases (alanine aminotransferase[ALT] and aspartate aminotransferase [AST]) and improves lipid profiles, such as cholesterol and triglyceride levels, in both children and adults. In a Phase III clinical trial (ARISE), patients receiving sebelipase alfa showed normalized ALT levels and reductions in hepatic fat fraction and liver volume [58]. Additionally, long-term studies have confirmed sustained benefits, with infants treated with sebelipase alfa showing higher survival rates and better management of hepatosplenomegaly and weight gain [59]. The safety profile of sebelipase alfa is generally favorable. The majority of adverse reactions are mild to moderate in severity, such as fever, nausea, abdominal pain, and allergies, which can be controlled with dose adjustments [58]. These results suggest that sebelipase alfa can be safely administered with proper monitoring, making it a viable long-term therapeutic option for managing LAL-D [52, 60]. Despite some challenges related to treatment costs and the need for long-term studies, its efficacy and safety have positioned it as a promising therapeutic option for managing this rare and debilitating disease.

Kanuma was eventually acquired by Alexion for a high price of $8.4 billion from Synageva BioPharm, the technology holder of Kanuma [61]. Using this method, it is not only possible to effectively produce large-scale recombinant therapeutic proteins, but the expression in egg white also ensures protein stability and activity, providing an innovative treatment solution for LAL-D. This success clearly demonstrates that transgenic chickens have vast development potential in the biopharmaceutical production field [6].

5. Regulatory, ethical, and biosafety considerations

5.1 Animal welfare and ethical frameworks

Genetic modification techniques provide an effective means for the genetic improvement of chickens, allowing for targeted enhancement of disease resistance. For example, editing genes such as ANP32A and TVA can confer resistance to avian influenza and avian leukosis [62, 63]. However, such modifications can also affect the health and behavior of the chickens. For instance, knocking out the TVA gene leads to abnormalities in vitamin B_{12} metabolism and elevated methylmalonic acid levels [64]. Studies have also shown that chickens with a functional loss of the PMEL17 gene exhibit less aggressive behavior and a stronger tendency to explore [65]. Several studies indicate that chimeric chickens and their offspring obtained using CRISPR technology are capable of producing target proteins, but their offspring exhibit lower production efficiency than wild-type [6]. Furthermore, CRISPR's off-target events should not be overlooked, including unintended mutations, gene deletions, genomic structural variations, and activation of protooncogenes, all of which present potential health and safety challenges for genetically edited animals [6].

The 3Rs principle (replacement, reduction, refinement) serves as the core guideline for animal welfare in experiments. In terms of replacement, researchers have developed long-term cultured chicken tracheal organoids, which efficiently support avian influenza virus replication and are used to assess the efficacy of antiviral drugs, reducing the use of live chickens by more than 90% [66]. In reduction, researchers have achieved efficient gene editing through PGCs transplantation or direct sperm manipulation, eliminating the need to keep large numbers of G0 generation chimeras until adulthood, thus reducing the breeding generations from three to two and decreasing experimental chicken usage by over 50% [67]. In optimization, researchers have introduced comprehensive enrichment facilities, such as perches, sand baths, pecking materials, and visual barriers, into the broiler rearing environment, which reduced stereotypical pacing by 65%, feather-pecking behavior by 50%, and improved the stability of experimental data by over 20% [68].

5.2 Environmental containment and biosafety

The environmental risks of genetically edited organisms and the core of biosafety lie in the dual isolation strategy, which combines both physical and biological isolation, along with the management of gene flow. The dual isolation strategy serves as the basic protective measure. Physical isolation blocks biological escape through the construction of multi-level enclosed breeding facilities [69]. Biological isolation, on the other hand, uses techniques such as reproductive isolation and genetic firewalls to prevent gene flow [70]. The potential risks of gene flow should be managed based on standardized systems, focusing on three core aspects: receptor species, transmission pathways, and ecological impact [71]. In practice, there are still very few cases that genetically modified animals can refer to, but the lessons from genetically modified plants are worth noting. For example, the gene flow from genetically modified maize in Mexico's center of origin led to the contamination of native maize germplasm, causing ecological imbalances [72]. Control measures can include physical isolation, such as geographical separation and adjusting flowering periods. Biological isolation, such as reproductive isolation and genetic firewalls, can further strengthen protection [70]. More exploration is needed for genetically modified animals, but it is essential to consider biosafety measures at the outset of their widespread application.

5.3 Regulatory pathways (FDA, EMA perspectives)

The FDA classifies genetically edited animals as inheritable genomic modifications (IGAs) and implements a three-tier, risk-based approval system under GFI #187A. Low-risk cases can undergo simplified reviews, while high-risk cases require the submission of complete safety and efficacy data [73]. The EMA, led by the CVMP, conducts centralized reviews and, under EU Regulation 2019/6, offers preconsultation and accelerated pathways, emphasizing environmental and food safety. The final approval of evaluation results is made by the European Commission [69].

The fate of cutting-edge gene-editing technologies is ultimately closely tied to societal acceptance. Currently, international regulations regarding genetically modified animals still need improvement, and many members of the public remain skeptical of genome-edited animals and their derived food products. Therefore, improvements at the legislative and administrative levels are necessary to enhance the supervision and authorization procedures for genetically modified livestock, as well as their related foodstuffs and drugs [6]. Transparent communication strategies, timely dissemination of the technology's potential benefits, known risks, and uncertainties, and the establishment of effective public engagement, consultation, and education mechanisms are essential. The successful market entry of products will be accelerated by these innovations, ultimately leading to broader application and increasing their market potential.

6. Future perspectives and conclusions

6.1 Integration with synthetic biology and artificial intelligence–driven design

The future development of chicken oviduct bioreactors is likely to be significantly accelerated by the combination of synthetic biology and artificial intelligence (AI)-driven design. Together, these provide powerful tools for enhancing the efficiency of genome engineering and biomanufacturing performance. The AI-based approaches, including deep learning predictions and protein language models, can significantly improve sgRNA design precision, discovery of new CRISPR/Cas variants, advanced gene network analysis, and more efficient donor delivery strategies, all of which will greatly improve the efficiency of targeted knock-ins [74]. Additionally, synthetic regulatory elements like oviduct-specific promoters, enhancers, and insulator expression boxes can be rationally designed to achieve higher and more stable expression levels in egg whites. Machine learning approaches leveraging large-scale genomic and transcriptomic datasets can further support predictions for optimal integration sites, regulatory structures, and protein yield outcomes, thereby advancing the precision and large-scale engineering of poultry production lines [6].

Beyond enhancing expression levels, synthetic biology also makes it possible to produce more complex therapeutic biologics in eggs. The new generation of biopharmaceuticals increasingly includes multimeric protein complexes, bispecific antibodies, and glycoengineered fusion proteins, which often require coordinated expression and complex posttranslational modifications. Using AI-driven de novo protein design allows for the creation of protein designs and synthesis based on functional structural modules, independent of known structural modules and evolutionary constraints [75]. Moreover, these AI-driven designs can be extended to enzymes, vaccines, antibodies, etc., and such designs inherently increase their functional unpredictability and complexity [76, 77]. The chicken oviduct system, combined with modular genetic circuit design, may become a unique platform to assemble such complex molecules in a cost-effective and scalable manner.

It is also noteworthy that the combination of rapid reproduction cycles and decentralized egg-based production models suggests that poultry bioreactors could play a role in emergency manufacturing. For example, with antimicrobial resistance, traditional antibiotic discovery methods struggle to keep up with the rapid mutations of viruses, creating an urgent need for new antibacterial treatments, especially during epidemic outbreaks. The synergy between AI and antimicrobial peptide (AMP) discovery can help us design safe and efficient new peptides, opening up new frontiers in the fight against resistance [78]. In future pandemic control or regional biological shortages, engineered poultry could support the distributed and flexible production of vaccines or therapeutic proteins, serving as an important supplement to centralized mammalian cell culture facilities.

6.2 Current technology status and global application prospects

For chicken bioreactors, the biggest challenge in development lies in selecting the technological pathway, which includes several aspects. First, therapeutic proteins or monoclonal antibody fragments are often large, and although viral vectors have high transgenic efficiency, the size of the foreign gene fragments they can deliver is limited. Developing vectors that can carry larger fragments with higher efficiency is urgent. Second, large-fragment random integration often impacts the health of chickens, and even genetically modified chickens tend to have reduced egg production performance. For reasons that remain unclear, eggs from genetically modified hens often display lower-than-expected levels of exogenous recombinant proteins in egg white. These eggs are also smaller than those of wild-type hens and contain comparatively less protein in the egg white [6]. Third, before the advent of CRISPR/Cas technology, the application of targeted gene insertion was limited to mice. Since the advent of CRISPR/Cas technology, significant progress has been made in targeted insertion in chickens. However, the integration of foreign fragments mainly occurs in ovalbumin or ovomucoid, which may inevitably affect the expression of endogenous genes. Therefore, identifying more suitable safe harbor sites for the targeted insertion of foreign genes is crucial for the advancement of chicken bioreactors.

In summary, chicken bioreactors express target genes in the oviduct, allowing the corresponding gene products to be secreted and accumulated in eggs. Although only a limited number of mature drugs are currently approved on the market, the oviduct bioreactor still demonstrates substantial potential in commercial production, basic research, and long-term sustainability. This production approach offers advantages such as high biosafety, low production costs, ease of purification, and protein modifications – including glycosylation – that more closely resemble their natural state. In recent years, the application and rapid development of advanced gene-editing technologies such as CRISPR/Cas9 have significantly improved the precision and efficiency of genetically modified chicken production, providing new possibilities for obtaining recombinant therapeutic proteins with ideal glycosylation patterns and lower immunogenicity. In the future, genetically modified chickens have the potential to completely transform the production methods of therapeutic proteins.

Author details

Lijuan Wang*, Yongkang Xie, Xue Wang, Qiang Zheng and Guojun Wang
Department of Basic Medicine and Forensic Medicine, Baotou Medical College, Baotou, China

*Address all correspondence to: wanglijuan-2002@163.com

References

[1] Jain S, et al. THPdb2: compilation of FDA approved therapeutic peptides and proteins. Drug Discovery Today. 2024;**29**(7):104047

[2] Walsh G, Walsh E. Biopharmaceutical benchmarks 2022. Nature Biotechnology. 2022;**40**(12):1722–1760

[3] Herron LR, et al. A chicken bioreactor for efficient production of functional cytokines. BMC Biotechnology 2018;**18**:82

[4] Sethuraman N, Stadheim TA. Challenges in therapeutic glycoprotein production. Current Opinion in Biotechnology. 2006;**17**(4):341–346

[5] Sheridan C. FDA approves 'farmaceutical' drug from transgenic chickens. Nat Biotechnol. 2016;**34**(2):117–119

[6] Meng S, et al. Genetically modified chickens as bioreactors for protein-based drugs. Front Genome Ed. 2025;**6**:1522837

[7] Kawabe Y, et al. Production of chimeric antibodies by transgenic chicken bioreactors. In Animal Cell Technology: Basic & Applied Aspects. Vol. **14**. Dordrecht: Springer Netherlands; 2006. p. 309–315

[8] Beyze A, Larroque C, Quintrec ML. The role of antibody glycosylation in autoimmune and alloimmune kidney diseases. Nature Rev Nephrol. 2024;**20**(10):672–689

[9] Li P, et al. Protein Glycosylation: An Emerging Regulator of Allergic Diseases. 2025;**0**:1–15

[10] Grewal US, Kurzrock R. Mucin-1: a promising pan-cancer therapeutic target. npj Precision Oncology. 2025;**9**(1):218

[11] Zhu L, et al. Production of human monoclonal antibody in eggs of chimeric chickens. Nat Biotechnol 2005;**23**:1159–1169

[12] Polejaeva IA. 25th Anniversary of cloning by somatic cell nuclear transfer: Generation of genetically engineered livestock using somatic cell nuclear transfer. Reproduction. 2021;**162**(1):F11–F22

[13] Perry MM. A complete culture system for the chick embryo. Nature. 1988;**331**(6151):70–72

[14] Love J, et al. Transgenic birds by DNA microinjection. Bio/technology (Nature Publishing Company) 1994;**12**(1):60–63

[15] Salter DW, et al. Gene insertion into the chicken germ line by retroviruses. Poultry Science 1986;**65**(8):1445–1458

[16] Harvey AJ, et al. Expression of exogenous protein in the egg white of transgenic chickens. Nature Biotechnology. 2002;**20**(4):396–399

[17] Uhm SJ, et al. Effect of Transgene Introduction and Recloning on Efficiency of Porcine Transgenic Cloned Embryo Production In Vitro.

Reproduction in Domestic Animals 2009;**44**:106–115

[18] Williams RM, Sauka-Spengler T. Ex ovo electroporation of early chicken embryos. STAR Protocols. 2021;**2** (2):100424

[19] Véron N, et al. CRISPR mediated somatic cell genome engineering in the chicken. Developmental Biology 2015;**407**(1):68–74

[20] Sid H, Schusser B. Applications of Gene Editing in Chickens: A New Era Is on the Horizon. Frontiers in Genetics. 2018;**9**:456

[21] International Chicken Genome Sequencing Consortium. Sequence and comparative analysis of the chicken genome provide unique perspectives on vertebrate evolution. Nature. 2004;**432** (7018):695–716

[22] Ezaki R, et al. Transcription activator-like effector nuclease-mediated deletion safely eliminates the major egg allergen ovomucoid in chickens. Food and Chemical Toxicology. 2023;**175**:113703

[23] Higashitani Y, Horie K. Long-read sequence analysis of MMEJ-mediated CRISPR genome editing reveals complex on-target vector insertions that may escape standard PCR-based quality control. Scientific Reports. 2023;**13**(1): 11652

[24] Chojnacka-Puchta L, Sawicka D. CRISPR/Cas9 gene editing in a chicken model: current approaches and applications. J Appl Genetics. 2020;**61** (2):221–229

[25] Oishi I, et al. Targeted mutagenesis in chicken using CRISPR/Cas9 system. Scientific Reports. 2016;**6**:23980

[26] Wang L, et al. Enhancing Targeted Genomic DNA Editing in Chicken Cells Using the CRISPR/Cas9 System. PLoS One 2017;**12**:e0169768

[27] Jenkins K, et al. Production of immune receptor knockout chickens via direct in vivo transfection of primordial germ cells. Animal biotechnology. 2025;**36**(1):2523027

[28] Whyte J, et al. FGF, Insulin, and SMAD Signaling Cooperate for Avian Primordial Germ Cell Self-Renewal. Stem Cell Reports 2015;**5**:1171–1182

[29] Han JY, Lee BR. Isolation and Characterization of Chicken Primordial Germ Cells and Their Application in Transgenesis. Methods Mol Biol. 2017;**1650**:229–242

[30] Bednarczyk M, et al. Generation of transgenic chickens by the non-viral, cell-based method: effectiveness of some elements of this strategy. J Appl Genetics 2018;**59**:81–89

[31] van de Lavoir MC, et al. Germline transmission of genetically modified primordial germ cells. Nature 2006;**441**:766–769

[32] Harel-Markowitz E, et al. Use of sperm plasmid DNA lipofection combined with REMI (restriction enzyme-mediated insertion) for production of transgenic chickens expressing eGFP (enhanced green fluorescent protein) or human follicle-stimulating hormone. Official journal of the Society for the Study of Reproduction 2009;**80**:1046–1052

[33] Cooper CA, et al. Generation of gene edited birds in one generation using sperm transfection assisted gene editing (STAGE). Transgenic Res. 2017;**26**:331–347

[34] Trefil P, et al. Preparation of fowl testes as recipient organs to germ-line chimeras by means of gamma-radiation. Br. Poult. Sci 2003;**44**:643–650

[35] Trefil P, et al. Restoration of spermatogenesis and male fertility by transplantation of dispersed testicular cells in the chicken. Official journal of the Society for the Study of Reproduction 2006;**75**:575–581

[36] Mucksová J, et al. Conservation of chicken male germline by orthotopic transplantation of primordial germ cells from genetically distant donors†. Official journal of the Society for the Study of Reproduction 2019;**101**:200–207

[37] Sakuma T, et al. MMEJ-assisted gene knock-in using TALENs and CRISPR-Cas9 with the PITCh systems. Nat. Protoc. 2016;**11**:118–133

[38] Ji KE, et al. CRISPR-Cas9: a promising tool for gene editing on induced pluripotent stem cells. Korean J Intern Med. 2017;**32**(1):42–61

[39] Urban W, et al. CRISPR-Cas9 in the Tailoring of Genetically Engineered Animals. Curr Issues Mol. Biol. 2025;**47** (5):330

[40] Rezazade Bazaz M, Dehghani H. From DNA break repair pathways to CRISPR/Cas-mediated gene knock-in methods. Life sciences. 2022;**295**:120409

[41] Cox DB, Platt RJ, Zhang F. Therapeutic genome editing: prospects and challenges. Nat. Med. 2015;**21** (2):121–131

[42] Yuan M, et al. HMEJ-based safe-harbor genome editing enables efficient generation of cattle with increased resistance to tuberculosis. Journal Biol Chem. 2021;**296**:100497

[43] Suzuki K, et al. In vivo genome editing via CRISPR/Cas9 mediated homology-independent targeted integration. Nature 2016;**540**:144–149

[44] Xie L, et al. HMEJ-mediated efficient site-specific gene integration in chicken cells. J. Biol. Eng. 2019;**13**:90

[45] Oishi I, et al. Efficient production of human interferon beta in the white of eggs from ovalbumin gene-targeted hens. Scientific Reports. 2018;**8** (1):10203

[46] Taleei R, Nikjoo H. Biochemical DSB-repair model for mammalian cells in G1 and early S phases of the cell cycle. Mutat Res Mutat Res. 2013;**756** (1–2):206–212

[47] Nayga J, et al. In Vitro Culture of Avian Primordial Germ Cells: Established Methods and Future Directions. Biology (Basel). 2025;**14** (11):1597

[48] Ichikawa K, et al. Knock-in of the duck retinoic acid-inducible gene I (RIG-I) into the Mx gene in DF-1 cells enables both stable and immune response-dependent RIG-I expression. Biochem Biophys Reports. 2021;**27**:101084

[49] Schusser B, et al. Expression of heavy chain-only antibodies can support B-cell development in light chain knockout chickens. Eur. J. Immunol 2016;**46**:2137–2148

[50] Qin X, et al. Efficient knock-in at the chicken ovalbumin locus using adenovirus as a CRISPR/Cas9 delivery system. 3 Biotech 2019;**9**:454

[51] Park TS, et al. Deposition of bioactive human epidermal growth factor in the egg white of transgenic hens using an oviduct-specific minisynthetic promoter. FASEB Journal 2015;**29**:2386–2396

[52] Shirley M. Sebelipase alfa: first global approval. Drugs. 2015;**75** (16):1935–1940

[53] Mukae T, Yoshii K, Oishi I. Purification of Recombinant Monoclonal Antibodies from Transgenic Chicken Eggs and Removal of Egg-White Proteins. Biol Pharma Bull. 2025;**48**(6):838–842

[54] Woodcock ME, et al. Reviving rare chicken breeds using genetically engineered sterility in surrogate host birds. P Natl Acad Sci USA 2019;**116**:20930–20937

[55] Macdonald J, et al. Characterisation and germline transmission of cultured avian primordial germ cells. PLoS One 2010;**5**:e15518

[56] Schusser B, et al. Immunoglobulin knockout chickens via efficient homologous recombination in primordial germ cells. P Natl Acad Sci USA 2013;**110**:20170–20175

[57] Zhao R, et al. Production of viable chicken by allogeneic transplantation of primordial germ cells induced from somatic cells. Nature Communications 2021;**12**:2989

[58] Bashir A, Tiwari P, Duseja A. Enzyme replacement therapy in lysosomal acid lipase deficiency (LAL-D): a systematic literature review. Ther Adv Rare Dis. 2021;**2**:26330040211026928

[59] Sebelipase alfa (Kanuma) for lysosomal acid lipase deficiency. Med Lett Drugs Ther. 2016, **58** (1504), e126–7

[60] Demaret T, et al. Sebelipase alfa enzyme replacement therapy in Wolman disease: a nationwide cohort with up to ten years of follow-up. Orphanet J. Rare Dis. 2021;**16**:507

[61] Ratner M. Alexion pays big for Synageva's rare disease drug candidate. Nat Biotechnol. 2015;**33**(7):679

[62] Idoko-Akoh A, et al. Creating resistance to avian influenza infection through genome editing of the ANP32 gene family. Nature Communications 2023;**14**:6136

[63] Hellmich R, et al. Acquiring Resistance Against a Retroviral Infection via CRISPR/Cas9 Targeted Genome Editing in a Commercial Chicken Line. Front Genome Ed. 2020;**2**:3

[64] Koslova A, et al. Knock-Out of Retrovirus Receptor Gene Tva in the Chicken Confers Resistance to Avian Leukosis Virus Subgroups A and K and Affects Cobalamin (Vitamin B (12))-Dependent Level of Methylmalonic Acid. Viruses. 2021;**13**(12):2504

[65] Karlsson AC, et al. Genotype on the pigmentation regulating PMEL17 gene affects behavior in chickens raised without physical contact with conspecifics. Behav. Genet 2011;**41**: 312–322

[66] de Bruin ACM, et al. Long-term culture of chicken tracheal organoids for

the purpose of avian influenza virus research. Virol J 2025;**22**:99

[67] Lee J, et al. Generation of genome-edited chicken and duck lines by adenovirus-mediated in vivo genome editing. P Natl Acad Sci USA 2022;**119**: e2214344119

[68] Guinebretiere M, et al. Structuring and enriching the rearing environment in conventional broiler chicken production: effects on behavioral indicators, emotional states, and cecal microbiota composition. Poultry sci. 2025;**104**(11):105663

[69] Organisms EPOGM, et al. New developments in biotechnology applied to animals: An assessment of the adequacy and sufficiency of current EFSA guidance for animal risk assessment. EFSA Journal 2025;**23**:e9566

[70] Wray-Cahen D, Hallerman E, Tizard M. Global regulatory policies for animal biotechnology: overview, opportunities and challenges. Front Genome Ed. 2024;**6**:1467080

[71] Philips JG, Martin-Avila E, Robold AV. Horizontal gene transfer from genetically modified plants - Regulatory considerations. Frontiers in Bioengineering and Biotechnology. 2022;**10**:971402

[72] Quist D, Chapela IH. Transgenic DNA introgressed into traditional maize landraces in Oaxaca. Mexico. Nature. 2001;**414**(6863):541–543

[73] Epstein LR, et al. CRISPR, animals, and FDA oversight: Building a path to success. P Natl Acad Sci USA. 2021;**118** (22):e2004831117

[74] Li Z, et al. From Code to Life: The AI-Driven Revolution in Genome Editing. Adv Sci (Weinh). 2025;**12**:e17029

[75] Zhang G, et al. Beyond evolution: De novo designed protein toolkit rewriting the rules of synthetic biology. Biosafety and Health 2025;7:306–311

[76] Olawade DB, et al. Leveraging artificial intelligence in vaccine development: A narrative review. Journal of Microbiological Methods. 2024;**224**:106998

[77] Cui XC, et al. AI-driven de novo enzyme design: Strategies, applications, and future prospects. Biotechnology Advance. 2025;**82**:108603

[78] Szymczak P, et al. AI-Driven Antimicrobial Peptide Discovery: Mining and Generation. Acc. Chem. Res. 2025;**58**:1831–1846

Chapter 3

Current Debate Regarding Laboratory and Natural Origin Hypotheses of SARS-CoV-2: A Critically Balanced Review

Ibrahim C. Haznedaroglu

Abstract

The emergence of Severe Acute Respiratory Syndrome Coronavirus 2 (SARS-CoV-2), leading to the COVID-19 syndrome, triggered a global pandemic with profound medical, social, and economically devastating consequences. The origin of SARS-CoV-2 yet remains one of the most critical scientific questions of our ongoing century. SARS-CoV-2 can be considered a genetically modified organism from the point of origin due to either a zoonotic spillover and/or a laboratory incident. Despite numerous scientific studies, no definitive evidence has resolved whether the virus emerged via a zoonotic spillover and/or started through viral research-related lab-leak modifications. This review follows a structured literature-search approach, integrates mutually supporting and opposing evidence for both hypotheses, outlines limitations, and aims to provide an objective synthesis for clinicians, molecular scientists, and regulatory experts. Current evidence does not conclusively support either a natural zoonotic spillover or a research-related laboratory incident. A cautious, evidence-based approach is essential to support global preparedness for future emerging pathogens.

Keywords: SARS-CoV-2 origin, sarbecovirus evolution, furin cleavage site (FCS), lab leak, biosafety, gain-of-function, restriction site analysis, Huanan market, ACE2 binding

1. Introduction and perspectives

The emergence of Severe Acute Respiratory Syndrome Coronavirus 2 (SARS-CoV-2) in Wuhan, China, prompted a global crisis, yet the precise origin of the virus remains a critical open scientific question [1]. SARS-CoV-2 can be considered a genetically modified organism from the point of origin due to either a zoonotic spillover and/or a laboratory incident. Although early investigations focused on the natural zoonotic origins [2], the possibility of a viral modification research-related lab-leak incident has persisted as well [3]. Five years after the initial viral outbreak, the overall scientific community lacks definitive evidence

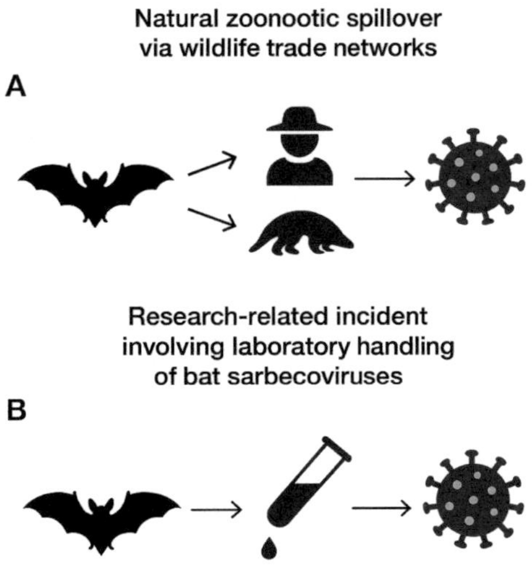

Figure 1.
Conceptual model illustrating two proposed pathways for the emergence of SARS-CoV-2. (a) Natural zoonotic spillover pathway: A schematic representation of wildlife-associated transmission routes through which a bat sarbecovirus may have entered the human population. The model includes ecological reservoirs (e.g., Rhinolophus bats), potential intermediate hosts involved in wildlife trade networks, and human exposure points such as live-animal markets. This pathway reflects known mechanisms for coronavirus emergence but remains unconfirmed due to the absence of a verified progenitor virus or identified intermediate host. (b) Research-related laboratory incident pathway: A schematic representation of a potential accidental exposure scenario involving field sampling, viral isolation, or experimental work with bat sarbecoviruses in a research laboratory setting. The model includes routine activities such as sample processing, animal experiments, and culture-based studies. This pathway is considered scientifically plausible but remains unconfirmed, as no direct evidence links SARS-CoV-2 to any reported laboratory virus, construct, or biosafety incident. Overall interpretation: Both pathways are presented as hypothetical and unresolved, reflecting current scientific uncertainty. The figure emphasizes that available data are insufficient to determine the origin of SARS-CoV-2 and that further ecological surveillance, epidemiological data access, and laboratory transparency are needed to assess these possibilities.

supporting either of the hypotheses [4] (**Figure 1**). Genomic sequencing confirms that SARS-CoV-2 is fundamentally a chimeric virus [2, 5, 6] well adapted for humans [7, 8]. The human body disease setting of the COVID-19 syndrome is well defined [9, 10]. SARS-CoV-2 exposure in humans is presented as three clinicobiological phases: first, the asymptomatic/pre-symptomatic phase; second, the respiratory phase with mild/moderate/severe symptoms; and third, the multi-systemic clinical syndrome with impaired/disproportionate and/or defective immunity [11]. The corresponding three immunogenomic markers are namely the ACE2 and ANPEP transcripts in the initial phase, EGFR and IGF2R transcripts in the propagating phase, and the immune system-related critical gene involvements of the complicating phase [9]. The SARS-CoV-2 chimera represented a great challenge to the human immune system as a strong "non-natural" "non-self" pathological antigenic stimulus, particularly at the initial induction phase of the pandemic process at the community level [12]. These kinds of clinical human observations regarding the rapid viral genesis cast doubt that, unlike typical zoonotic spillover events, SARS-CoV-2 possesses specific molecular features and environmental paradoxes that suggest an artificial acquisition of high virulence and human adaptation [13].

The aim of this review is to provide a balanced assessment of both hypotheses [4], focusing on the biological features relevant to zoonotic spillover and laboratory methodologies, based on the current virological, genomic, epidemiological, and biosafety literature. A structured literature search was conducted in PubMed, Scopus, Web of Science, and medRxiv (preprints interpreted with caution). Search terms included SARS-CoV-2 origin, sarbecovirus evolution, furin cleavage site (FCS), serial passage, lab leak, biosafety, gain-of-function, restriction site analysis, Huanan market, ACE2 binding, and related keywords. The search inclusion criteria were peer-reviewed publications (2020–2025), peer-reviewed modeling studies, high-impact genomic, phylogenetic, biosafety, and structural biology analyses, as well as official reports from WHO, NIH, and related health authorities. This perspective is intended to provide a neutral and evidence-based synthesis of current scientific knowledge relevant to both hypotheses. The exclusion criteria for the analyses were speculative non-scientific commentary, non-transparent secondary sources, and/or politically motivated documents without methodological rigor.

Given the persistent lack of definitive evidence for either scenario, a rigorous scientific approach requires ongoing assessment of both hypotheses. This includes careful consideration of uncertainties, avoidance of speculative claims, and a commitment to transparent scientific discourse. Clarifying the origin of SARS-CoV-2 remains important not only for historical understanding but also for guiding global practices in wildlife surveillance, biosafety, and oversight of research involving potential pandemic pathogens.

2. Overview of the origin hypotheses

2.1 Natural zoonotic spillover hypothesis

The natural zoonotic spillover hypothesis proposes that SARS-CoV-2 originated from wildlife, likely through a sarbecovirus, potentially involving one or more intermediate hosts before infecting humans (**Table 1**, **Figure 1**). This mechanism is consistent with the emergence of SARS-CoV-1, MERS-CoV, and other coronaviruses that crossed species barriers. Thus, the main principle is that SARS-CoV-2 arose from an un-sampled bat sarbecovirus, potentially through an intermediate host, before infecting humans. Evidence supporting this hypothesis indicates that SARS-CoV-1 and MERS-CoV emerged via zoonotic pathways; wet-market animal trading networks have historically linked wildlife to humans; early infected human cases clustered spatially around the Huanan Seafood Market; and several researchers identified environmental SARS-CoV-2–positive samples associated with commercial wildlife stands [14–17]. However, several key gaps still remain in the zoonotic spillover hypothesis of the SARS-CoV-2 virus. Despite the supporting experiments and theoretical basis, substantial uncertainties remain. No animal reservoir or intermediate host has been confirmed; no direct progenitor virus with >99% genomic identity has been discovered; and wildlife surveillance is incomplete across large geographic regions. While these gaps do not negate the spillover hypothesis, they highlight the need for expanded ecological sampling and greater availability of early case data. Therefore, a research laboratory-related viral leak incident cannot be ruled out despite extensive scientific discussions on that sensitive matter [7, 14, 15, 18–25].

Feature	Description in SARS-CoV-2	"Zoonotic spillover" interpretation	"Laboratory-modifications" interpretation
Genomic backbone and mosaic structure	SARS-CoV-2 shares ~96% genome identity with RaTG13 (bat CoV), while its RBD region shares similarity with pangolin CoVs.	Coronaviruses frequently recombine in bat and wildlife hosts; mosaic genomes are common in sarbecoviruses due to co-infection and template switching.	Chimeric viruses combining backbone and RBD regions can also be produced in reverse-genetics systems during studies of host range and pathogenesis.
Spike RBD affinity for human ACE2	SARS-CoV-2 exhibits high binding affinity for hACE2.	High-affinity binding may arise through natural selection in wildlife hosts or early human circulation, as seen in many emerging viruses.	Serial passage in cell culture or animal models can select for enhanced ACE2 binding, though no direct evidence links SARS-CoV-2 to such adaptations.
FCS at S1/S2 (PRRA insertion)	Unique polybasic FCS has not yet been identified in known sarbecoviruses but is present in other coronavirus lineages.	Insertions and cleavage sites can arise through natural mechanisms, such as recombination, polymerase slippage, or template switching.	Polybasic sites can be inserted experimentally using reverse-genetics; FCS insertions have been proposed in some research contexts. No direct evidence connects the SARS-CoV-2 FCS to such work.
Codon usage in PRRA insertion (CGG–CGG)	Contains a rare arginine doublet codon pair among known bat CoVs.	Rare codons can appear via stochastic insertions or recombination. CGG codons exist naturally elsewhere in coronaviruses.	CGG codons are used in mammalian expression systems and can appear in engineered sequences. Their presence alone is not proof of laboratory manipulation.
Restriction enzyme site patterns (Type IIS sites)	Some analyses have noted spacing patterns compatible with modular cloning; others find such patterns common in natural viruses.	Random evolution can produce evenly spaced restriction sites; simulations suggest patterns fall within natural variability.	Type IIS sites may be used in synthetic assembly systems; theoretical compatibility has been discussed but lacks direct supporting evidence for SARS-CoV-2.
Human adaptation features	SARS-CoV-2 replicates efficiently in human airway cells and exhibits broad tissue tropism.	Adaptation may occur naturally before or after spillover; early variants also show increased adaptation, indicating ongoing natural evolution.	Pre-adaptation could theoretically occur through laboratory passaging in human or humanized models; no known experimental record supports this.

Table 1.
Key genomic and biological features of SARS-CoV-2 and their alternative interpretations.

2.2 Research-related lab-leak incident hypothesis

The research-related incident hypothesis proposes that SARS-CoV-2 may have entered the human population through an accidental laboratory exposure involving field sampling, viral culture, or research on bat sarbecoviruses (**Table 1, Figure 1**). This hypothesis does not require intentional genetic engineering; it encompasses accidental infection during viral isolation, experimental passaging, or handling of uncharacterized viral samples. The comprehensive analysis of the unique viral genomic structure and the contextual research environment reveals multiple anomalies that necessitate the thorough examination of laboratory modifications as a likely pathway of emergence. Unlike typical zoonotic spillover events, SARS-CoV-2 possesses specific molecular features and environmental paradoxes that suggest a non-natural acquisition of high virulence and human adaptation [13–15, 18]. Biosafety lapses in BSL-2 settings are historically documented across multiple countries. The lab-leak hypothesis indicates an accidental community infection initiated due to laboratory research work on sarbecoviruses with the aim of producing a "pan-coronavirus vaccine." Laboratory work mainly included viral isolation and passaging, field sampling activities, and handling of unpublished viral strains. Supporting the lab-leak hypothesis involves contextual elements: Wuhan houses the world's largest known collection of bat sarbecoviruses, the already present scientific research regarding chimeric coronaviruses, RBD swapping, and humanized mouse models [1, 26, 27]. Directed lab-research could evolve coronaviruses for the study of pathogenesis and host range expansion. The analyses of key pathobiological features suggestive of laboratory modification in SARS-CoV-2 are depicted in **Table 1**. However, several key gaps still remain in the lab-leak hypothesis of the SARS-CoV-2 virus [28, 29]. First, no direct link has been established between SARS-CoV-2 and any known laboratory virus; and second, no publicly released database records confirm possession of a near-progenitor strain. Therefore, a natural zoonotic spillover origin cannot be ruled out despite extensive scientific discussions on this sensitive matter [2, 30–32]. Thus, neither of these two hypotheses meets the standard of scientific conclusive proof [4]. No direct evidence links SARS-CoV-2 to any known laboratory virus, reverse-genetics construct, or published research sequence. No near-progenitor virus has been reported in any laboratory database, and no documented biosafety breach has been shown to involve SARS-CoV-2 or a closely related virus. As with the spillover hypothesis, the research-related incident hypothesis remains scientifically plausible but unproven [2, 30–32].

2.3 Current state of evidence

At present, neither hypothesis satisfies the evidentiary standards required for a definitive conclusion. Both pathways are consistent with some available data, both face notable evidentiary gaps, and both require further investigation. A scientifically neutral approach acknowledges this uncertainty and recognizes that distinguishing between these hypotheses will depend on expanded wildlife sampling, improved transparency regarding early-case data, and accessible laboratory records relating to coronavirus research activities [2, 30–32].

3. Genomic features of SARS-CoV-2: Neutral presentation of evidence

Genomic analyses have played a central role in evaluating hypotheses concerning the origin of SARS-CoV-2. Several features of the viral genome – such as receptor-binding characteristics, the FCS, codon usage, and restriction-site patterns – have been discussed in both natural and laboratory-origin scenarios. None of these features alone provides definitive evidence for either hypothesis, and each must be interpreted within the broader context of coronavirus evolution and research methodologies [2, 30–32].

3.1 Spike receptor-binding domain (RBD) and ACE2 affinity

Genomic sequencing confirms that SARS-CoV-2 is fundamentally a chimeric virus, binding to the human ACE2 with high affinity [5, 33, 34]. The numerous upcoming variants of the virus (Alpha, Delta, Omicron, and many others) independently evolved even stronger ACE2 binding. The binding selection of the spike RBD could support natural adaptability but does not exclude pre-adaptation mechanisms such as serial passage in laboratory models [4, 13, 15]. The RBD of the SARS-CoV-2 Spike protein displays a level of pre-adaptation that is difficult to justify by natural selection pathways leading directly to humans. The vast majority of its genome sequence shares 96.2% identity with the bat coronavirus RaTG13 (renamed from BtCoV/4991) [35]. The critical genetic divergence occurs in the RBD of the Spike protein. This RBD region is almost identical to that found in a pangolin coronavirus, specifically the MP789 strain. The resulting SARS-CoV-2 genome, therefore, appears to be a composite of a RaTG13-like backbone and a pangolin-like RBD. For such a precise genetic chimerism to occur naturally, the two distinct precursor viruses must infect the same host cell simultaneously, facilitating a recombination event [36]. The subsequent recovery of the infectious chimeric virus Bat-SRBD, such as a Bat-SCoV backbone plus SARS-CoV RBD, indicated the functional interchangeability of the core components and explicitly identified the Spike RBD as the key determinant of cross-species transmission. The chimeric genomic structure, with the components sourced from geographically disparate hosts like bat and pangolin, demands consideration of laboratory modification since chimeric viruses can arise via both natural recombination and lab-research [4, 14, 15, 18]. Natural spillover typically involves an initial period of inefficient, low-affinity binding, followed by the rapid accumulation of mutations within the human host to optimize binding. The immediate, high-affinity binding of SARS-CoV-2 to human ACE2, coupled with its highly efficient replication in human cells compared to bat cells, is consistent with the direct introduction of a virus that was already pre-adapted through prolonged selection in a research laboratory environment simulating human infection. The subsequent recovery of the infectious chimeric virus Bat-SRBD, such as a Bat-SCoV backbone plus SARS-CoV RBD, indicated the functional interchangeability of the core components and explicitly identified the Spike RBD as the key determinant of cross-species transmission [4, 14, 15, 18]. At present, the RBD sequence and its binding affinity are compatible with both natural recombination and laboratory adaptation scenarios. Current data do not allow discrimination between these hypotheses.

3.2 The furin cleavage site

The polybasic FCS at the Spike S1/S2 junction is arguably the most compelling genomic signature [37]. SARS-CoV-2 is the only one of more than 800 known SARS-related coronaviruses (sarbecoviruses) to possess an FCS. The FCS is unique within known sarbecoviruses but not within coronaviruses broadly, although similar motifs exist in HKU1, OC43, MERS-CoV, and other lineages [38]. The FCS is associated with significantly increased viral transmissibility and virulence. There are two interpretations for the existence of the FCS in SARS-CoV-2, in accordance with the natural zoonotic origin and/or a viral modification research-related

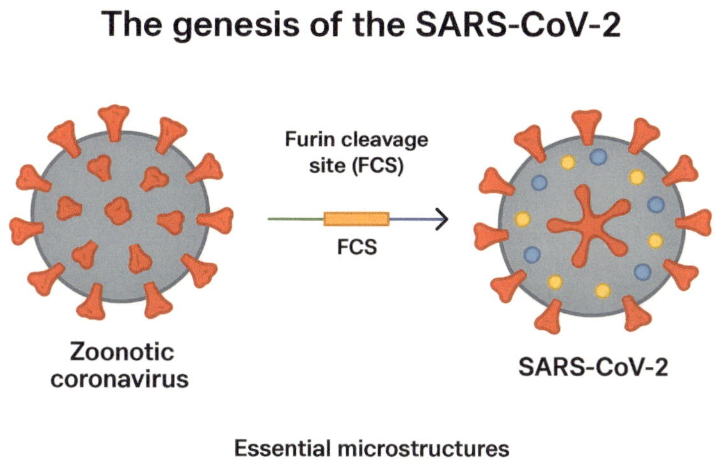

Figure 2.
The existence of a polybasic FCS, which is associated with significantly increased viral transmissibility and virulence, at the spike S1/S2 junction of SARS-CoV-2. SARS-CoV-2 is the only one of the SARS-related coronaviruses (sarbecoviruses) to possess an FCS. There are two interpretations for the genesis of the FCS in SARS-CoV-2, in accordance with the natural zoonotic spillover origin and/or viral modification research-related lab-leak incident hypotheses.

lab-leak incident hypotheses (**Figure 2**). First, the natural acquisition of the FCS by the virus via recombination hotspots near S1/S2; and second, the insertion within a laboratory research setting through reverse genetics [18, 38–40]. Thus, current evidence does not permit definitive attribution [41, 42]. Laboratory modification involves numerous techniques like reverse genetics and synthetic genomics, which facilitate the assembly and modification of viral components, such as inserting a new RBD or FCS. Likewise, experimental adaptation and site-directed mutagenesis could rely on applying strong selective pressure through prolonged serial passage in highly permissive systems, such as human airway epithelial cell cultures and/or humanized mouse models. This process selects for increased virulence and optimal host binding affinity [7]. Current data neither confirm nor exclude natural acquisition or laboratory modification of the FCS. Its evolutionary origin remains uncertain.

Retention of the FCS and the full length of the spike protein [43] in the design of mRNA-spike medications rarely caused a syndrome called "Spikeopathy" in a sub-population of vaccinated people [44]. The synthetic spike retains the natural furin cleavage site (PRRAR) at the S1/S2 junction. This site is crucial for proteolytic processing during viral entry. However, in the vaccine context, cleavage by host proteases does not facilitate infection but may aid in antigen presentation. The benefits of COVID-19 vaccines, such as preventing severe disease, decreasing hospitalization, and reducing death, far outweigh potential risks, but hepatic constitutive sustained spike synthesis is critical for the genesis of Spikeopathy syndrome in a human subpopulation [45–47]. The hypothetical mechanistic clinicopathological components of Spikeopathy include genomic instability, immune escape, altered immune surveillance, impaired DNA repair mechanisms, destabilization of the regulation of cell growth and apoptosis, chronic inflammation, immune system dysregulation, Renin-angiotensin system (RAS) dysregulation, activation of oncogenic pathways, and alteration of the tumor microenvironment via accumulation of lipid nanoparticles [45, 46, 48].

3.3 Codon usage and insert sequence

The FCS functional site was created by a precise 12-nucleotide insertion (TCCTCGGCGGGC) encoding the PRRA amino acid sequence. SARS-CoV-2 does have a cleavage site in the Spike protein, and it also contains two arginine amino acids, identified by the triplet CGG. This triplet, in fact, is common in human cells but is rarely found in wild bat CoVs. The CGG-CGG arginine codon pair is particularly rare in Sarbecoviruses. The insertion enables cleavage by the ubiquitous host-cell protease furin. Since furin is expressed broadly across various human organs and tissues, the FCS confers the virus with the unique ability to cross species and tissue barriers, leading to a systemic disease profile. There are two interpretations for the existence of FCS in SARS-CoV-2 in accordance with the natural zoonotic origin and/or viral modification research-related lab-leak incident hypotheses [4, 14, 15, 18]. First, the synthetic acquisition of CGG within laboratory research; and second, a natural zoonotic spillover could produce the CGG-CGG arginine codon pair since codon usage is heterogeneous across coronaviruses, and natural insertions may lead to rare motifs, as CGG appears elsewhere in CoV genomes [18, 38–40]. The presence of the human-like, non-bat-like CGG codon usage within the uniquely inserted FCS is considered a molecular footprint providing strong support for a research lab-related origin [49]. Overall, the codon usage pattern of the FCS insertion remains a point of debate but does not provide definitive evidence for either hypothesis.

3.4 Restriction site patterns

The groundbreaking work relied on combining phylogenetic and bioinformatics analysis with large-scale cDNA synthesis, enabling the construction and transcription of a full-length infectious cDNA genome based entirely on hypothetical sequences, bypassing the need to successfully cultivate primary isolates from nature. The SARS-CoV-2 genome also exhibits physical evidence of its artificial assembly process. The genome structure displays a pattern of regularly spaced Type IIS restriction endonuclease sites that is indicative of assembly from six distinct synthetic nucleic-acid building blocks using the enzyme BsmBI. The use of BsmBI is characteristic of seamless directional cloning, a common method in synthetic genomics. Moreover, the FauI restriction site is embedded within the FCS sequence itself, a feature that could have facilitated cloning, restriction fragment length polymorphism screening, or mutation tracking. The simultaneous co-existence of five unique, signature genomic features, including optimized RBD, unique FCS, rare CGG codon usage, BsmBI assembly signature, and the resulting high human-specific pre-adaptation, could indicate the laboratory modification of the SARS-CoV-2 virus [50]. On the contrary, Type IIS restriction patterns like BsmBI have been opposed by several analyses, implying that random viral evolution could generate similar patterns, and simulations indicated the frequent occurrence of evenly spaced restriction sites [51]. As with other features, restriction-site distribution has multiple plausible explanations, and current data do not establish whether these patterns reflect natural evolution or laboratory methodology.

3.4.1 Summary of genomic evidence

All major genomic features of SARS-CoV-2 can be interpreted under both natural and laboratory-related frameworks. None constitutes conclusive evidence for either origin, and each interpretation is subject to limitations, such as incomplete viral sampling in nature, limited access to laboratory records, and uncertainties in reconstructing historical viral evolution.

4. Evidence relevant to natural spillover

Analyses of the earliest documented COVID-19 cases suggest a significant spatial association with the Huanan Seafood Wholesale Market (HSM) in Wuhan. Multiple independent studies have reported that many of the earliest cases – particularly those without known links to other confirmed infections – clustered around the HSM. Two early viral lineages (A and B) were identified among cases near the market, which some researchers interpret as being compatible with multiple spillover events or rapid early spread in a crowded market environment. Retrospective mapping has indicated that early cases not directly linked to the market were still geographically concentrated around the surrounding district, suggesting this area may have played a role in early amplification [17, 52]. While this evidence is consistent with zoonotic transmission, it does not conclusively identify the market as the site of initial introduction, and the precise epidemiological sequence remains unresolved due to incomplete early-case clinical data and limited access to initial patient records. Environmental sampling conducted in early 2020 identified SARS-CoV-2 RNA on surfaces in multiple stalls associated with live-animal trade. Positive samples were found near stalls selling wildlife species known to carry other coronaviruses, such as raccoon dogs, red foxes, and other mammals historically implicated in zoonotic disease transmission. These environmental detections do not constitute evidence of infected animals but suggest that the virus circulated in areas where live animals were kept or sold. Wildlife trade networks in China and neighboring regions include species susceptible to sarbecoviruses. Databases, market records, and wildlife trafficking assessments have documented the legal and informal movement of animals capable of hosting related viruses. However, no animal sampled at the HSM tested positive for SARS-CoV-2. Viral RNA found in environmental samples does not distinguish between human-to-environment contamination and environment-to-human or animal-to-human pathways. Thus, environmental evidence is supportive but not decisive [53–55].

Multiple studies have already indicated early case clustering around the Huanan Market. Spatial mapping suggested that at least two early lineages were circulating near the market. Pangolin-CoVs and bat-CoVs share partial features with SARS-CoV-2 but lack the required similarity to constitute direct progenitors. Likewise, environmental swabs from wildlife-associated stalls tested positive for SARS-CoV-2 RNA [17, 52]. Furthermore, the illegal or informal wildlife trade is well documented in Wuhan. However, there are significant gaps in evidence since no infected animal has been found, and the precise transmission pathway remains unconfirmed. No bat virus with sufficient genome similarity has been identified as a proximal progenitor, and critically, no natural intermediate host capable of facilitating the leap to humans has been found [53–55].

The natural spillover hypothesis proposes that SARS-CoV-2 emerged through transmission from wildlife to humans, potentially involving one or more intermediate host species. This pathway is consistent with the emergence of previous coronaviruses, including SARS-CoV-1 and MERS-CoV. Although no definitive progenitor or intermediate host has yet been identified, several lines of epidemiological and ecological evidence support the plausibility of a wildlife-associated origin. At the same time, notable gaps remain that prevent firm conclusions.

Despite supporting scientific evidence, the natural-origin hypothesis remains unconfirmed due to key missing elements:

1. Absence of a confirmed progenitor virus with >99% genome identity.

2. No animal reservoir or intermediate host identified to date.

3. Incomplete wildlife surveillance, particularly in regions where related bat coronaviruses circulate.

4. Limited availability of early-case data, which constrains reconstruction of initial transmission events.

These limitations do not refute the spillover hypothesis but underscore the need for continued surveillance and transparent access to epidemiological, ecological, and sequencing data.

5. Evidence relevant to a research-related incident

The research-related laboratory incident hypothesis encompasses several possible pathways through which SARS-CoV-2 could have entered the human population, including accidental infection during field sampling, viral isolation, laboratory passaging, or work involving genetic manipulation. Importantly, this hypothesis does not assume intentional engineering; it includes accidental exposure to naturally occurring or modified viruses under study. While this scenario remains scientifically plausible, no direct evidence currently links SARS-CoV-2 to any specific laboratory activity, virus strain, or research construct. At the same time, certain contextual factors motivate continued examination of this hypothesis.

5.1 Published laboratory capabilities

The reverse genetics systems for the lab-research of the sarbecoviruses were well established even during the time period between 2008 and 2019 [56]. Mouse-adapted SARS-CoV-2 variants could be generated through serial passage [13–15, 18]. There are clues that chimeric spike swapping had been performed within some research work (e.g., SHC014-CoV, WIV1-CoV studies) [57, 58]. Naturally, those data could demonstrate only the feasibility of the lab-leak hypothesis and not direct proof. The setting of conspiracy theories is established via the documentation of the NIH Grant AI110964 (2014–2019) and DARPA DEFUSE (2018) research projects, mainly focusing on the construction of novel chimeras of coronaviruses with enhanced ability to infect and replicate in human cells.

5.2 Biosafety considerations regarding the virus-research laboratories

There may be global biosafety gaps, as some coronavirus work historically occurred in BSL-2 settings in research laboratories, leading to the accidental escape of the virus into the community. Performing research involving genetically enhanced pathogens with potentially massive transmissibility under BSL-2 conditions poses a high risk of infection for laboratory staff. Past lab-acquired SARS infections have demonstrated real-life risks [59]. Likewise, there are significant data transparency limitations, as access to raw viral databases remains restricted, and metadata associated with early cases and sampling logs remains incomplete. These gaps limit the definitive exclusion of a laboratory research-related incident as the genesis of SARS-CoV-2. Given these constraints, biosafety considerations support the plausibility of accidental exposure but do not provide direct evidence of such an event.

Key uncertainties prevent confirmation of a lab-research-related origin:

1. No identified laboratory virus or construct with sufficient genomic similarity to SARS-CoV-2.

2. No documented laboratory exposure or breach linked to a SARS-like virus in late 2019.

3. No publicly available records demonstrating that the specific combination of genomic features in SARS-CoV-2 existed in any laboratory strain.

4. Incomplete access to laboratory logs, metadata, and field-sampling records limiting independent verification.

These unresolved issues leave the laboratory-incident hypothesis scientifically plausible but unproven.

6. Conclusions, perspectives, and recommendations

Determining the origin of SARS-CoV-2 remains an unresolved scientific challenge. Both leading hypotheses – natural zoonotic spillover and a research-related laboratory incident – remain plausible but unconfirmed. Current genomic, epidemiological, ecological, and biosafety evidence provides important insights, yet substantial gaps in data prevent definitive attribution to either pathway. A balanced interpretation acknowledges these uncertainties and emphasizes the need for continued, evidence-driven investigation rather than premature conclusions. An integrated synthesis of any objective review focusing on the SARS-CoV-2 origin must indicate persistent current scientific uncertainty, as both zoonotic spillover and laboratory-related incident hypotheses are plausible but unconfirmed [4, 60, 61]. SARS-CoV-2's origin remains unresolved [62, 63]. Genomic features can be interpreted under both natural and laboratory scenarios, as incomplete access to primary data prevent resolution [64]. Both hypotheses retain scientific plausibility but are constrained by substantial evidentiary gaps [3, 65, 66]. Genomic, structural, epidemiological, and biosafety data provide partial and sometimes conflicting insights. The balanced interpretation presented here aligns with scientific requirements and highlights the pressing need for transparency, rigorous biosafety, and robust international collaboration. Only through improved data access and global oversight can such questions be conclusively answered and future pandemics mitigated [23, 24]. A responsible scientific position acknowledges the asymmetrical data gaps rather than definitive, conclusive claims [2, 67]. The main limitations of the current evidence include, firstly, the absence of the progenitor virus preventing firm conclusions; secondly, the deficiency of key data regarding patient zero identification, lab sample inventories, and wildlife sampling coverage. Thus, both hypotheses rely on indirect inference, and political constraints have limited independent scientific origin investigations [68].

Meanwhile, there must be essential scientific policy and research recommendations that shall be established within the context of both viral origin hypotheses [59, 69]. High-containment requirements should be strengthened. Research on potential pandemic pathogens (PPP) and gain-of-function involving enhanced transmissibility shall occur only in BSL-3/4 laboratory settings. Laboratory incident reporting systems must be improved. Wildlife sampling may be expanded, but the immediate release of viral sequence data from surveillance programs should be mandatory. Public registration of high-risk pathogen research protocols consistent with international biosafety guidance could also be established to prevent similar uncertainty in future pandemics.

Genomic features of SARS-CoV-2, including receptor-binding characteristics, the FCS, and codon and restriction-site patterns, can be interpreted within both natural evolutionary and laboratory contexts. None of these features serves as conclusive proof for either hypothesis. Similarly, epidemiological analyses show early case associations with wildlife trade environments but do not identify an infected animal host. Laboratory research on related sarbecoviruses demonstrates the technical feasibility of potential laboratory pathways, yet no direct evidence links SARS-CoV -2 to any known laboratory strain, construct, or incident.

A responsible scientific perspective requires transparent acknowledgment of evidentiary limitations:

- Absence of a confirmed progenitor virus in wildlife or laboratory collections.
- No identified intermediate host, despite extensive sampling.
- Limited access to early case data, clinical records, and environmental samples.
- Restricted availability of laboratory inventories and research logs relevant to coronavirus studies.

These gaps underscore the importance of strengthened transparency and international collaboration. Regardless of which hypothesis ultimately proves correct, or whether the true origin remains indeterminate, several policy and research recommendations are broadly applicable. Research involving potential pandemic pathogens, including reverse-genetics systems and viral passaging, should be conducted under stringent containment, preferably BSL-3/4, with robust oversight, training, and incident-reporting mechanisms. Harmonizing international biosafety frameworks would reduce risks associated with high-consequence pathogen research. Expanded genomic surveillance of bats and other potential reservoir species, along with improved monitoring of wildlife-trade networks, is essential to detect precursor viruses and understand pathways of viral emergence. Timely and transparent release of genomic sequences from surveillance programs should be prioritized. Independent scientific inquiry requires open access to early outbreak data, environmental sampling results, viral sequences, and laboratory inventories. Global initiatives enabling secure, standardized data sharing – while protecting sensitive information – would facilitate objective assessment of competing hypotheses. Joint fieldwork, laboratory partnerships, and coordinated data analyses involving international teams can strengthen trust and improve the quality and reproducibility of origin-related research. Collaborative platforms may help resolve key uncertainties and support global preparedness. The origin question is complex and sensitive. Its investigation must remain firmly rooted in evidence-based scientific inquiry rather than political narratives or speculative claims. Emphasizing transparency and methodological rigor is essential to avoid bias and foster productive debate.

Author details

Ibrahim C. Haznedaroglu
Hacettepe University Medical School, Department of Internal Medicine and Hematology, Ankara, Turkey

*Address all correspondence to: ichaznedaroglu@gmail.com

References

[1] Zhou P, Yang X-L, Wang X-G, Hu B, Zhang L, Zhang W, et al. A pneumonia outbreak associated with a new coronavirus of probable bat origin. Nature 2020;**579**:270–273

[2] Andersen KG, Rambaut A, Lipkin WI, Holmes EC, Garry RF. The proximal origin of SARS-CoV-2. Nature Medicine. 2020;**26**(4):450–452

[3] Bloom JD, Chan YA, Baric RS, Bjorkman PJ, Cobey S, Deverman BE, et al. Investigate the origins of COVID-19. Science. 2021;**372**:694

[4] Ruiz-Medina BE, Varela-Ramirez A, Kirken RA, Robles-Escajeda E. The SARS-CoV-2 origin dilemma: Zoonotic transfer or laboratory leak? BioEssays. 2022;**44**(1):2100189

[5] Turk C, Turk S, Temirci ES, Malkan UY, Haznedaroglu IC. In vitro analysis of the renin-angiotensin system and inflammatory gene transcripts in human bronchial epithelial cells after infection with severe acute respiratory syndrome coronavirus. Journal of the Renin-Angiotensin-Aldosterone System. 2020;**21**(2):1470320320928872

[6] Hou YJ, Okuda K, Edwards CE, Martinez DR, Asakura T, Dinnon KH, et al. SARS-CoV-2 reverse genetics reveals a variable infection gradient in the respiratory tract. Cell 2020;**182**: 429–46.e14

[7] Zhan SH, Deverman BE, Chan YA. SARS-CoV-2 is well adapted for humans. What does this mean for re-emergence?. bioRxiv. 2020. DOI: 10.1101/2020.05.01.073262

[8] Massey SE. Host manipulation mechanisms of SARS-CoV-2. Acta Biotheoretica. 2022;**70**(1):4

[9] Turk C, Turk S, Malkan UY, Haznedaroglu IC. Three critical clinicobiological phases of the human SARS-associated coronavirus infections. European Review for Medical and Pharmacological Sciences. 2020;**24**:8606–8620

[10] Rabaan AA, Smajlović S, Tombuloglu H, Ćordić S, Hajdarević A, Kudić N, et al. SARS-CoV-2 infection and multi-organ system damage: A review. Biomolecules and Biomedicine. 2023;**23**:37–52

[11] Malkan UY, Haznedaroglu IC. Hematological aspects of the COVID-19 syndrome. European Review for Medical and Pharmacological Sciences. 2022;**26**(12):4463–4476

[12] Turk S, Turk C, Malkan UY, Temirci ES, Peker MC, Haznedaroglu IC. Current community transmission and future perspectives on the COVID-19 process. Turkish Journal of Medical Sciences. 2021;**51**(3): 1001–1011

[13] Segreto R, Deigin Y, McCairn K, Sousa A, Sirotkin D, Sirotkin K, et al. Should we discount the laboratory origin of COVID-19? Environmental Chemistry Letters. 2021;**19**: 2743–2757

[14] Deigin Y, Segreto R. The genetic structure of SARS-CoV-2 is consistent with both natural or laboratory origin: Response to Tyshkovskiy and Panchin. BioEssays. 2021;**43**(9):2100137

[15] Deigin Y, Segreto R. SARS-CoV-2's claimed natural origin is undermined by issues with genome sequences of its relative strains: Coronavirus sequences RaTG13, MP789 and RmYN02 raise multiple questions to be critically addressed by the scientific community. Bioessays. 2021;**43**(7):2100015

[16] Frutos R, Gavotte L, Devaux CA. Understanding the origin of COVID-19 requires to change the paradigm on zoonotic emergence from the spillover to the circulation model. Infection, Genetics and Evolution. 2021;**95**:104812

[17] Worobey M, Levy JI, Malpica Serrano L, Crits-Christoph A, Pekar JE, Goldstein SA, et al. The Huanan Seafood Wholesale Market in Wuhan was the early epicenter of the COVID-19 pandemic. Science. 2022;**377**:951–959

[18] Segreto R, Deigin Y. The genetic structure of SARS-CoV-2 does not rule out a laboratory origin: SARS-COV-2 chimeric structure and furin cleavage site might be the result of genetic manipulation. Bioessays. 2021;**43**(3):2000240

[19] Balaram P. Natural and unnatural history of the coronavirus. Current Science. 2021;**120**(12):1820–1826

[20] Gøtzsche PC. Made in China: The coronavirus that killed millions of people. Indian Journal of Medical Ethics. 2022;7:254

[21] Van Helden J, Butler CD, Achaz G, Canard B, Casane D, Claverie J-M, et al. An appeal for an objective, open, and transparent scientific debate about the origin of SARS-CoV-2. The Lancet 2021;**398**:1402–1404

[22] Yan L-M, Kang S, Guan J, Hu S. Unusual features of the SARS-CoV-2 genome suggesting sophisticated laboratory modification rather than natural evolution and delineation of its probable synthetic route. 2020. Preprint

[23] Harrison NL, Sachs JD. A call for an independent inquiry into the origin of the SARS-CoV-2 virus. Proceedings of the National Academy of Sciences. 2022;**119**(21):e2202769119

[24] Harrison NL, Sachs JD. Reply to Garry: The origin of SARS-CoV-2 remains unresolved. Proceedings of the National Academy of Sciences. 2022;**119**(45):e2215826119

[25] Decroly E, Claverie J-M, Canard B. The WHO mission report struggles to trace the origins of the SARS-CoV-2 epidemic. Virologie. 2021;**25**(3):42–46

[26] Leist SR, Dinnon KH, Schäfer A, Tse LV, Okuda K, Hou YJ, et al. A mouse-adapted SARS-CoV-2 induces acute lung injury and mortality in standard laboratory mice. Cell. 2020;**183**:1070–85.e12

[27] Menachery VD, Yount BL, Debbink K, Agnihothram S, Gralinski LE, Plante JA, et al. A SARS-like cluster of circulating bat coronaviruses shows potential for human emergence. Nature Medicine. 2015;**21**:1508–1513

[28] Coccia M. Meta-analysis to explain unknown causes of the origins of SARS-COV-2. Environmental Research. 2022;**211**:113062

[29] Domingo JL. What we know and what we need to know about the origin of SARS-CoV-2. Environmental Research. 2021;**200**:111785

[30] Worobey M, Levy JI, Serrano LMM, Crits-Christoph A,

Pekar JE, Goldstein SA, et al. The Huanan market was the epicenter of SARS-CoV-2 emergence. Science 2022;**377**(6609):951–959

[31] Frutos R, Javelle E, Barberot C, Gavotte L, Tissot-DuPont H, Devaux CA. Origin of COVID-19: Dismissing the Mojiang mine theory and the laboratory accident narrative. Environmental Research. 2022;**204**:112141

[32] Boni MF, Lemey P, Jiang X, Lam TT-Y, Perry BW, Castoe TA, et al. Evolutionary origins of the SARS-CoV-2 sarbecovirus lineage responsible for the COVID-19 pandemic. Nature Microbiology. 2020;**5**:1408–1417

[33] Aladag E, Tas Z, Ozdemir BS, Akbaba TH, Akpinar MG, Goker H, et al. Human Ace D/I polymorphism could affect the clinicobiological course of COVID-19. Journal of the Renin-Angiotensin-Aldosterone System. 2021;**2021**:5509280

[34] Haznedaroglu IC. Immunogenomic phases of COVID-19 and appropriate clinical management. Lancet Microbe. 2020;**1**(7):e278

[35] Massey SE. Forensic genomic analysis determines that RaTG13 was likely generated from a bat mating plug. Microbiology Research. 2024;**15**(3):1784–1805

[36] Akaishi T. Insertion-and-deletion mutations between the genomes of SARS-CoV, SARS-CoV-2, and bat coronavirus RaTG13. Microbiology Spectrum. 2022;**10**(3):e00716–22

[37] Argyrou M, Pitsillou E, Hung A, El-Osta A, Karagiannis TC. Insights into the pathogenic mechanisms associated with the SARS-CoV-2 spike protein. Journal of Structural Biology. 2025;**217**(3):108229

[38] Chan YA, Zhan SH. The emergence of the spike furin cleavage site in SARS-CoV-2. Molecular Biology and Evolution. 2021;**39**(1):1–7. DOI: 10.1093/molbev/msab327

[39] Nguyen HT, Zhang S, Wang Q, Anang S, Wang J, Ding H, et al. Spike glycoprotein and host cell determinants of SARS-CoV-2 entry and cytopathic effects. Journal of Virology. 2021;**95**(5). DOI: 10.1128/jvi.02304-20

[40] Harrison NL, Sachs JD. What is a furin cleavage site, why is it important, and how might this have arisen in SARS-CoV-2. Zenodo. 2023;**1.0.**:1–32

[41] Romeu AR. On-going evidence on SARS-CoV-2 furin site arginine codon usage bias and new found PRRA-like insertions. Europe PMC Plus. 2023;**686691**:1–28. DOI: 10.21203/rs.3.rs-2748884/v1

[42] Romeu AR, Ollé E. SARS-CoV-2 and the secret of the furin site. Europe PMC Plus. 2021;**20944**:1–17. DOI: 10.20944/preprints202102.0264.v1

[43] Morais P, Adachi H, Yu Y-T. The critical contribution of pseudouridine to mRNA COVID-19 vaccines. Frontiers in Cell and Developmental Biology. 2021;**9**:2021

[44] Parry PI, Lefringhausen A, Turni C, Neil CJ, Cosford R, Hudson NJ, et al. 'Spikeopathy': Covid-19 spike protein is pathogenic, from both virus and vaccine mRNA. Biomedicines 2023;**11**:2287

[45] Erdoğdu B, Kaplan O, Fidan BB, Çelebier M, Malkan ÜY,

Haznedaroglu IC. Metabolomic profiling of leukemic hematopoiesis: Effects of BNT162b2 mRNA COVID-19 vaccine administration. Current Molecular Medicine. 2025. DOI: 10.2174/0115665240361878250601074746

[46] Erdoğdu B, Kaplan O, Çelebier M, Malkan ÜY, Haznedaroğlu İC. Hepatic metabolic dysregulation as a potential amplifier of leukemogenesis following mRNA vaccination: A novel mechanistic hypothesis. Medicina. 2025;**61**(9):1687

[47] Yong SJ, Kenny TA, Halim A, Munipalli B, Alhashem YN, AlSaihati H, et al. Post-COVID-19 Vaccination (or Long Vax) syndrome: Putative manifestation, pathophysiology, and therapeutic options. Reviews in Medical Virology. 2025;**35**(5):e70070

[48] Cinar OE, Erdogdu B, Karadeniz M, Unal S, Malkan UY, Goker H, et al. Comment on Zamfir et al. Hematologic malignancies diagnosed in the Context of the mRNA COVID-19 vaccination campaign: A report of two cases. Medicina. 2022;**58**:1575

[49] Deigin Y, Segreto R. SARS-CoV-2's claimed natural origin is undermined by issues with genome sequences of its relative strains: Coronavirus sequences RaTG13, MP789 and RmYN02 raise multiple questions to be critically addressed by the scientific community. Bioessays. 2021;**43**(7):e2100015

[50] Bruttel V, Washburne A, VanDongen A. Endonuclease fingerprint indicates a synthetic origin of SARS-CoV-2. BioRxiv. 2022;**2022**:10.18.512756

[51] Wu F. Updated analysis to reject the laboratory-engineering hypothesis of SARS-CoV-2. Environmental Research. 2023;**224**:115481

[52] Crits-Christoph A, Levy JI, Pekar JE, Goldstein SA, Singh R, Hensel Z, et al. Genetic tracing of market wildlife and viruses at the epicenter of the COVID-19 pandemic. Cell. 2024;**187**:5468–82.e11

[53] Dinnon III KH, Leist SR, Schäfer A, Edwards CE, Martinez DR, Montgomery SA, et al. A mouse-adapted model of SARS-CoV-2 to test COVID-19 countermeasures. Nature. 2020;**586**:560–566

[54] Temmam S, Vongphayloth K, Baquero E, Munier S, Bonomi M, Regnault B, et al. Bat coronaviruses related to SARS-CoV-2 and infectious for human cells. Nature. 2022;**604**:330–336

[55] Xiao K, Zhai J, Feng Y, Zhou N, Zhang X, Zou J-J, et al. Isolation of SARS-CoV-2-related coronavirus from Malayan pangolins. Nature. 2020;**583**:286–289

[56] Becker MM, Graham RL, Donaldson EF, Rockx B, Sims AC, Sheahan T, et al. Synthetic recombinant bat SARS-like coronavirus is infectious in cultured cells and in mice. Proceedings of the National Academy of Sciences. 2008;**105**:19944–19949

[57] Hauser BM, Sangesland M, St Denis KJ, Lam EC, Case JB, Windsor IW, et al. Rationally designed immunogens enable immune focusing following SARS-CoV-2 spike imprinting. Cell Reports. 2022;**38**:110561

[58] Tse AL, Lasso G, Berrigan J, McLellan JS, Chandran K, Miller EH. Bat sarbecovirus WIV1-CoV bears an

adaptive mutation that alters spike dynamics and enhances ACE2 binding. PLOS Pathogens. 2025;**21**(10):e1013123

[59] Epstein GL. The Evolution of United States Governance Policies for Research Using Pathogens with Enhanced Pandemic Potential (ePPPs). Applied Biosafety. 2025;**30**(2):79–96

[60] Débarre F, Hensel Z. Theories of the origin of SARS-CoV-2 in the light of its continuing evolution. Comptes Rendus. Biologies. 2025;**348**:189–209

[61] Hao YJ, YL W, MY W, Zhou L, Shi JY, Cao JM, et al. The origins of COVID-19 pandemic: A brief overview. Transboundary and Emerging Diseases. 2022;**69**:3181–3197

[62] Yoo J-H. On the controversies surrounding the lab-leak theory of COVID-19. Journal of Korean Medical Science. 2025;**40**(16):1–3. DOI: 10.3346/jkms.2025.40.e153

[63] Thakur N, Das S, Kumar S, Maurya VK, Dhama K, Paweska JT, et al. Tracing the origin of Severe acute respiratory syndrome coronavirus-2 (SARS-CoV-2): A systematic review and narrative synthesis. Journal of Medical Virology. 2022;**94**:5766–5779

[64] Flegr J, Zahradník D, Zemková M. Thus spoke peptides: SARS-CoV-2 spike gene evolved in humans and then shortly in rats while the rest of its genome in horseshoe bats and then in treeshrews. Communicative & Integrative Biology. 2022;**15**(1):96–104

[65] Courtier-Orgogozo V, Bahulikar RA, Butler CD, Claverie J-M, Decroly E, Dudley JP, et al. SARS-CoV-2 origin is still unknown. Science. 2022:**337**(6609):951–959

[66] Pekar JE, Magee A, Parker E, Moshiri N, Izhikevich K, Havens JL, et al. The molecular epidemiology of multiple zoonotic origins of SARS-CoV-2. Science. 2022;**377**:960–966

[67] Ebright RH, MacIntyre R, Dudley JP, Butler CD, Goffinet A, Hammond E, et al. Implementing governmental oversight of enhanced potential pandemic pathogen research. Journal of Virology. 2024;**98**(4):e0023724

[68] Débarre F, Hensel Z. Theories of the origin of SARS-CoV-2 in the light of its continuing evolution. Comptes Rendus. Biologies. 2025;**348**(G1):189–209

[69] Lloyd Jr Jr RM, Hunter JR, Diaz RS, Schechter M. Could well-established, widely available, and simple laboratory techniques explain a laboratory origin of SARS-CoV-2? Brazilian Journal of Infectious Diseases. 2023;**27**(5):102808

Section 2

Ethical Considerations

Chapter 4

Perspective Chapter: Ethical Issues Associated with Genetically Modified Crops, and Can They Be Alleviated?

John R. Caradus and James A. Turner

Abstract

Ethics is about conforming to a standard of what is perceived as right and good for the community and the world at large. One reason why people reject science is that the information or achievement contradicts what they believe is true, good, or valuable. This has been the case with the advent of genetic modification in the mid-1990s, where concerns center on potential risks and benefits, including issues of environmental impact, human health, food safety, corporate control, social justice, and the concept of "unnaturalness." This review does not question the ethical standards of researchers, which require honesty, objectivity, and precision in all aspects of research, but rather the conflict of belief systems regarding what is good and right for individuals, society, and the environment. It will attempt to provide arguments to assist in aligning scientific advancements with mainstream and marginal societal values. Ethical concerns of plant research, and in particular plant biotechnology, are discussed. Simple answers and universal agreement are unlikely, but the debate on ethical concerns of genetic modification should recognize scientific evidence while ensuring that any engagement is centered on understanding the differences in human well-being, environmental impacts, and social justice that matter to different groups.

Keywords: belief systems, ethics, genetic modification, moral philosophy, science achievements, societal values, trust

1. Introduction

1.1 Defining what ethics is and is not

Ethics, also referred to as "moral philosophy" is defined as the discipline or conduct concerned with the concepts of what is morally good and right or bad and wrong [1]. The system or code that defines these concepts may be moral rules, principles, or values, which may encompass "religions, cultures, professions, or virtually any other group that is at least partly characterised by its moral outlook".

For an individual, ethics provides a set of standards by which decisions are made and activities can be judged as right or wrong [2].

Ethics are not the same as feelings, religion, following the law, or following accepted norms (i.e., a set of standards by which a particular group or community decides to regulate its behavior) [2, 3]. As the [4] stated, "one important difference between an ethical and a legal assessment framework is that the criteria are often not crystallised or 'established.' This means that an ethical assessment will have to be far more flexible, dynamic and open than a legal assessment, and allow more room for plurality." Additionally, while science can provide important information to assist with making informed ethical choices, science alone is not ethics. Therefore, while some activities may be scientifically or technologically possible, they may be considered by some individuals or groups to be unethical to develop and deploy [2, 5]. Ethical decision-making therefore considers adopting "a set of standards by which a particular group or community decides to regulate its behaviour – to distinguish what is legitimate or acceptable in pursuit of their aims from what is not" [6]. In this way, ethical principles provide the basis for certain key questions to be asked before decisions are made [7]. It has been argued that "a central core of universal values that any truly modern society must possess, and these are very much the values that science promotes, include rationality, creativity, the search for truth, adherence to codes of behaviour and a certain constructive subversiveness" [8]. Science is not value-neutral. It carries and promotes ethical ideals that are foundational to modern, just societies. This implies that science provides its own lens on ethical decision-making when undertaking a new initiative through prioritizing reason and evidence, promoting human benefit, upholding integrity and accountability, and encouraging critical engagement with societal norms [9].

Bioethics is an expression used to refer to ethics associated with research involving biology, biotechnology, environmental science, medicine, and agriculture [10]. It was originally used to indicate the "attempt to generate wisdom, the knowledge of how to use knowledge for social good from a realistic knowledge of man's biological nature and of the biological world" [11]. It was later used to mean the analysis of the benefits and risks arising out of the moral choices affecting living organisms for the good of individuals, the environment, and society – "the concept of love, balancing benefits and risks of choices and decisions" [12]. Bioethics is viewed in three ways – "(1) Descriptive bioethics is the way people view life, their moral interactions, and responsibilities with living organisms in their life; (2) Prescriptive bioethics is to tell others what is ethically good or bad, or what principles are most important in making such decisions; and (3) Interactive bioethics is discussion and debate between people, groups within society, and communities about 1 and 2 above". In summary, "the ideal is to avoiding doing harm and trying to do good" [12] and having a "consciousness of how our actions affect other species and the habitat which sustains us all" [8].

It has been proposed that "the quality of an ethical tool is determined by how well it achieves its assigned purpose(s)" [13]. "Ethical tools are designed with the aim of helping their users improve deliberations in order to reach ethically informed judgments or decisions." However, it is also understood that "tools will be used with various degrees of competence, and if a tool is used badly by a user, this does not mean that the tool is bad; however, a tool must allow for competent use in the context for which it was designed" [13]. Ethical tools have been placed into three categories by [13]:

1. Tools designed to support decision-making processes that involve multiple stakeholders;

2. Tools to foster public participation and dialogue around ethical issues; and

3. Tools used for ethical reflection or analysis are not necessarily tied to decision-making or public engagement.

The essential quality(s) needed for an ethical tool to be successfully used in delivering its purpose include [14]:

1. For all ethical tools, comprehensiveness and user-friendliness are crucial for the quality of the tool.

2. For tools that have reaching a decision as a main purpose, transparency, guiding users to a decision according to ambition, and justification of the decision-supporting mechanism are important qualities.

3. For tools that aim to engage the public, procedural fairness is essential.

Codes of ethics are often developed by professional organizations and are based on fundamental principles such as integrity, objectivity, professional competence and due care, confidentiality, and professionalism [15, 16]. One would assume that these principles would also be the basis for developing ethical decision frameworks. Similarly, ethical standards can be determined either prescriptively or interactively. Examples of processes that have been proposed to guide ethical decision-making are described in the following paragraphs. These vary from being highly interactive, as described by the "consensus conference" or citizen deliberation [17], to possibly a more prescriptive process that resulted in the principles proposed by [18].

An "Ethical Matrix Manual" was designed by [19] to "help decision-makers (as individuals or working in groups) reach sound judgments or decisions about the ethical acceptability and/or optimal regulatory controls for existing or prospective technologies in the field of food and agriculture." The matrix is made up of two dimensions – (1) the standard principles of respect for wellbeing, autonomy, and fairness; and (2) "interest groups" (i.e., affected parties) that are relevant to the issue in question. In the case of agricultural biotechnology, this might include farmers, food producers, and consumers, but also non-humans, such as farm animals. Use of the ethical matrix would "raise awareness of a wide range of ethical issues, encourage ethical reflection, provide a common basis for ethical decision-making, identify areas of agreement between individuals who might nevertheless differ in their overall judgments, clarify the basis of disagreements, and make explicit the reasoning that underpins any ethical decisions." However, despite the matrix potentially providing some clarity of thinking, "it is not possible to automatically arrive at a unique or prescribed course of action from the use of the ethical matrix." Case studies using this matrix have been published – for example, on genetically modified fish [20] or genetically modified canola [21].

The "consensus conference," is another ethical tool that involves panels of about a dozen or so selected laypersons who deliberate, with the help of experts, on a particular controversy, typically concerning some technology [22]. With the main objective being to involve the public in decision-making and coming to a consensus, it therefore requires people to balance different considerations and assign them relative weights.

A "Framework for Ethical Decision Making" has been recently published to assist individuals and groups in making "carefully considered decisions that will ultimately be more congruent with their own choices and values" [23]. This framework employs six ethical lenses to assist with deciding what is right or wrong, good or bad. These lenses are:

1. The rights lens – resulting in a decision or action that best protects and respects the moral rights of those affected.

2. The justice lens – to ensure that people are given fair and equal treatment.

3. The utilitarian lens – uses a results-based approach that leads to a decision or action producing the greatest balance of good over harm for as many stakeholders as possible.

4. The common good lens – suggests that the interlocking relationships of society are the basis of ethical reasoning and that respect and compassion for all others are paramount.

5. The virtue lens – determines that ethical actions ought to be consistent with certain ideal virtues that provide for the full development of our humanity. These virtues include honesty, courage, compassion, generosity, tolerance, love, fidelity, integrity, fairness, self-control, and prudence.

6. The care ethics lens – seeks to listen and respond to individuals in their specific circumstances rather than merely following rules or calculating utility.

In making an ethics-based decision, first, the ethical issues need to be identified. Then, information and facts relevant to the situation need to be gathered, followed by using the six lenses to decide how to act or respond. Through using the different lenses, groups and individuals can surface hidden values and assumptions, balance competing priorities (e.g., rights vs. outcomes), and make decisions that are more ethically robust and personally congruent [5].

Principles that might guide ethical evaluation have been outlined by [18] as:

- Scientific rigor and transparency are achieved through robust, independent scientific assessment of GM crops.

- Consider the precautionary principle, which ensures thorough risk assessment before wide deployment and monitoring after release.

- Social and economic impact assessment to understand the broader social, economic, and cultural implications.

- Meaningful choice through labeling of genetically modified foods, despite scientific institutions maintaining that approved genetically modified foods are safe, allows consumers to select according to their own risk assessments and values.

1.2 Aim of this review

The aim here is to examine the conflict of belief systems regarding what is good and right for individuals and society as it applies to plant research, and in particular, plant biotechnology research. It is not questioning the ethical standards of researchers, which require honesty, objectivity, and precision in all aspects of research. It will attempt to provide arguments to assist in aligning scientific advancements with both mainstream and marginal societal values. The questions to be examined here are:

- What are the different ethical concerns within society toward plant breeding, plant biotechnology, and genetic modification?
- How can "ethical tools" be used to discuss and respond to these concerns?

It has been observed [14] that "the polarization of the public debate stems from normative conflicts, rather than a lack of empirical research." A normative conflict occurs when two or more ethical norms or values clash, making it difficult to determine the most ethical course of action. These conflicts are not just disagreements about facts but more about what ought to be done (or not done). Additionally, there is an accepted view, and indeed a result of rational thinking, that "not everything that is technically feasible is ethically desirable" [5, 8].

1.3 Genetic modification of plants

Plant breeding has been defined as "any method of plant population reproduction that results in a permanent, heritable change" [24]. Humankind has been modifying plants of agricultural, horticultural, and ornamental value through a variety of plant breeding methods for over 10,000 years [25–32]. Indeed, it has been stated that "the crossing and the selection applied to the plants growing along the centuries have led to changing the genetic structure of plants, so that we could consider that nearly all species of cultivated plants are genetically modified" [33]. However, it has been acknowledged that "during the past 50 years, no other branch of biology has developed as fast and comprehensively as genetics, which analyses the inheritance and development" of living organisms [32]. Plant breeding and biotechnology have been major beneficiaries of this increased understanding of genetics. The benefits, risks, intended, and unintended consequences associated with the genetic modification of plants have been reviewed extensively [34, 35].

Over time, and certainly more recently, plant breeding methods have become more precise and controlled, with a better knowledge of expected results. What began as "selection by experience" [32] has progressed through the selection of targeted traits and controlled crossing, to the use of chemical and radiation-induced mutation to increase potential variability, to using tissue and cell culture techniques to rescue wide crosses, to hybrid breeding to allow the expression of heterosis (i.e., hybrid vigor), the use of polyploidy to overcome the non-viability and infertility of interspecific hybrids, obtain seedless polyploid cultivars, and increase resistance/tolerance to biotic and abiotic factors, through to using either Agrobacterium or biolistics to insert genes of known function into genomes of important crop plants (classically referred to as genetic modification or genetic engineering), and finally to new breeding techniques (which include gene editing), where this process is more targeted and precise

[32, 36–40]. In addition to genetic modification being shown to improve on-farm profitability and production [41] and being used to increase resistance against insects, herbicides, diseases, and abiotic stresses [42–45], it also has the potential to be used to deliver crops with improved nutrient quality, biofuel production, and enhanced production of vaccines and antibodies [46, 47].

As plant breeding processes have become more precise and accurate, both the level of regulation and cost, and time needed to obtain freedom to operate have increased considerably [48, 49, 50, 51, 52]. Interestingly, this has occurred not just for agriculturally important plants but also for genetically modified ornamental cultivars and has led to the conclusion that "the most significant barrier to market is the difficulty of managing, and the high cost of obtaining, regulatory approval" [53]. However, recently, the use of some forms of gene editing, which do not involve introducing foreign DNA, has resulted in a reduction in regulatory oversight in some jurisdictions [54–56]. But again, the global landscape here is variable and uncoordinated [54, 57]. Ideally, regulatory systems should be risk-appropriate, requiring evidence-based arguments to ensure decision-making is efficient and appropriate and, as such, does not stifle innovation [58].

Plant biotechnology research is defined here as an activity that combines genetics, molecular biology, and plant science to enhance agricultural practices and develop new plant cultivars with desirable traits [59]. Biotechnology research will result in the detection and mapping of genes, discovery of their functions, selection for specific genes in genetic resources and breeding, and transfer of genes for specific traits into plants where they are needed [60]. The areas of research required to achieve these outcomes include – understanding plant genetic structure and mechanisms, developing methods of genetic engineering or transfer of genes between organisms, identification of traits and genes associated with those traits that would improve crop yields and quality while lessening their impact on the environment, and, most recently, gene editing. For clarity, gene editing is distinguished from genetic engineering or transgenics by manipulating or inserting genes with a more precise and targeted approach [47, 61, 62]. Gene editing uses site-directed nucleases (SDN), which can include mega nucleases, Zinc-Finger Nucleases, transcription activator-like effector nucleases (TALENs), and the CRISPR-Cas systems [63]. These can lead to one of three outcomes – gene disruption or deletion (SDN1), gene correction or modification (SDN2), or DNA insertion (SDN3) [64, 65]. In a number of countries, SDN1 gene edits are unregulated since no foreign DNA has been introduced into the genome [66]. Indeed, it has been recognized that "CRISPR-Cas9 technology introduces new ethical and regulatory dilemmas related to its impact on health, the environment, and society" [67]. This relates to the relatively simple construction and low cost of CRISPR/Cas9 for genome editing, with the possibility of multiple purposes [68, 69], and the debate on whether some forms of gene editing are included or not within current gene technology regulations [70]. Additionally, a complex patent landscape "may negatively affect the development, research, and access to CRISPR techniques" [69].

2. Ethical permission requirements for plant research

Ethics in experimental research has been equated with scientific integrity and a moral responsibility to use acceptable and agreed-upon standards throughout the research process [71]. They stipulated that the "ethical code and policy include, but are not limited to, honesty, objectivity, integrity, carefulness, openness, intellectual rights recognition, confidentiality, responsible publication, social responsibility, competency, legality, and protecting research objects/subjects (plant, animal, human) from possible unfair manipulation."

Before genetic modification became commercially used, the concept of potentially unethical behavior in plant research was raised with regard to the use of indigenous knowledge about the medical benefits of endemic plants, which was then used to develop a dissertation, a scientific paper, or a drug [72]. The issue here is that often neither "native populations nor host countries have shared to a significant extent the financial benefits from any drug that reaches the market," which was viewed as running "the risk of serving ethically questionable purposes."

Plant biotechnology may be considered as either intrinsically wrong in itself and/or extrinsically wrong because of its consequences [7]. From an intrinsic viewpoint, the issues may result from "assumptions about the religious and moral status of the natural world and about the relationship that mankind should have with that world" [73]. As pointed out by Straughan (1995a), these cannot necessarily be proven right or wrong but can be analyzed by asking questions. "Should religious believers see modern biotechnology as a blasphemous affront to God or a creative opportunity to work with him? Can any moral guidelines be provided by what happens in Nature? Can any charges of intrinsic moral wrongness be brought against crop biotechnology which are not equally applicable to traditional methods of selective breeding?" Alternatively, "extrinsic concerns include issues such as the balance between risks and benefits, the ethics of patenting life-forms, and the likely socioeconomic impact of biotechnology" [73]. The primary concerns here are related to the mismanagement of intellectual property protection, resulting in "the monopolization of genetic resources, the inability to access life-saving medicines, and the ethical dilemmas posed by the patenting of living things" [74]. The balance to be ideally achieved is adequate protection of intellectual property to promote innovation, but with public interest as a priority to ensure accessibility through fair licensing policies.

A decade and a half ago, the European Commission released a document under the title of "Responsible Research and Innovation" (RRI) in an attempt to drive "a transparent, interactive process by which societal actors and innovators become mutually responsive to each other with a view on the (ethical) acceptability, sustainability, and societal desirability of the innovation process and its marketable products (in order to allow a proper embedding of scientific and technological advances in our society)" [75]. This treatise addresses a wide range of technological innovation opportunities from the viewpoint of "protection to the environment and human health, sustainability, and societal desirability". The concept of "responsible innovation" was recently explored by [76] through the development of a framework that integrated anticipation (considering potential impacts and implications of technologies), reflexivity (critically examining underlying assumptions and values), inclusion (engaging diverse stakeholders in decision-making), and responsiveness (adapting innovation pathways considering new knowledge and societal needs).

3. Ethical views of plant biotechnology

Reports published in 1999 and 2003 by the Nuffield Council on Bioethics [77, 78] identified the following ethical concerns about genetically modified crops – potential harm to human health, potential damage to the environment, negative impact on traditional farming practices, excessive corporate dominance, and the "unnaturalness" of the technology. These claims have been analyzed in light of the principles of general human welfare, the maintenance of human rights, and the principle of justice [79]. Other ethical concerns have included transparency, equitable access [59], and food sovereignty [18].

In a survey of 1,013 participants in the Netherlands, [80] indicated that 73.5% would support the use of genetic modification of wheat to remove gluten expression. The main concerns from those not supportive were – unacceptable health risks, negative long-term consequences for society primarily because it is unnatural, harm to the environment, and the "yuck-factor" (i.e., a feeling of horror, revulsion, or disgust) in that genetic modification would not cure the cause of the problem but merely its expression.

The Consultative Group on International Agricultural Research (CGIAR) organized a workshop in 1997, seeking to outline ethical and equity principles for the conservation and use of genetic resources [81]. The topic was approached from four perspectives, which are relevant to the purpose of the current review: (1) the scientific ethic, which underpins the quality and unique contribution of research in agricultural science, (2) a moral concern ultimately linked to the value of human life and the dignity of human beings, particularly as it relates to equity and poverty alleviation, (3) the ethical basis for the use of genetic resources to create new biological products that can transform and improve the lives of people, and (4) the ethics of conservation based on the value placed on the total biosphere as the best means to ensure continued progress and maintenance of the quality of human life. This workshop was partly motivated by concern over "whether scientists are acting in the best interests of people, including consumers of agricultural products which are the result of genetic engineering." This is potentially an example of applying some of the Markkula Centre six lenses for ethical decision-making [2, 23].

A review of 113 articles on moral reasoning associated with the use of genetically modified crops, expressed in academic publications from 1975 to 2008, identified that 84% addressed concerns about the consequences of the use of genetically modified crops, 57% addressed concerns about the technique itself, and 43% addressed the virtues related to the handling of risks and the application of genetically modified crops [82]. This indicated that there is no clear dichotomy between moral concerns addressing consequences and the act itself. The review also indicated that the moral reasoning of academic experts is well-founded on ecological concerns, as distinct from laypeople, whose concerns are more often associated with personal lifestyles and individual preferences [83–86]. This highlights that scientists are often using an ethical lens that prioritizes virtue, the common good, and utilitarianism, while the lay person uses an ethical lens that prioritizes rights and justice when considering the uses of genetic modification.

Discussion points for each of the identified ethical concerns are summarized below.

3.1 "Unnaturalness" of the technology: "playing god" and the sanctity of nature

The consideration that nature and naturalness are the preferred states compared to the management of the environment by humankind has been referred to as "the benevolence of nature" [87]. This has led to a strong preference for so-called natural and organic foods, particularly in developed countries [88–91]. However, outside of organics, naturalness is not a regulated condition, nor is there a shared or common understanding of what can be considered natural [14, 92]. Vega Rodriguez [87] describes some features that are affiliated with the naturalness of food, such that adding something to a so-called natural entity is considered to reduce naturalness, although removing items/components creates little concern; the greater the degree of processing, the less natural a food is considered; and, interestingly, domestication is not seen to negatively impact the perception of naturalness. Indeed, often naturalness is defined more by the absence of negative features, such as additives (no pesticides, no preservatives, and no chemicals) and processing (including no alterations, no contact with humans, and no industrial intervention) [93], which almost certainly includes genetic modification [94].

Genetic modification impacts can create "deep moral and religious connotations of the manipulation of the biological basis of life and of humankind" [95]. As a result, some consider that "biotechnology permits the direct control of the evolution of life, which many people believe is a usurpation of a divine role" [95]. Perhaps biotechnology might be considered more directed [93], but as mentioned earlier in this review, humankind has been influencing the evolutionary direction, particularly of crop plant species, for millennia using a variety of plant breeding methods. However, for some religious groups, moving animal genes into plant foods may be seen to violate dietary restrictions [96], and the use of appropriate labeling would be needed. Labeling has the dilemma that, in doing so, it may create a stigma about food from genetically modified crops, but the need for transparency is equally strong [97]. It has been argued that labeling may increase costs and curtail free and fair trade of genetically modified foods, and that food labels are there to provide consumer choice and can give the false impression that genetically modified food is less safe than conventional foods [98]. The tension between transparency and potential misinterpretation remains unresolved, and there is a view that disclosing the presence of genetically modified ingredients in food through labeling may be more misleading than informative [99]. Other authors [14] have argued that, while labeling enables individual's right to choose, it may not address some consumers' concerns about potential damage to ecosystems and social injustices.

A related criticism of agricultural biotechnology and, more specifically, genetically modified crops is a "virtue-based objection that to engage in agricultural biotechnology is to exhibit arrogance, hubris, and disaffection" [100]. Sandler explains that "while hubris does not justify opposition to all agricultural biotechnologies, it has merit and establishes a presumption against the use of biotechnology and GM crops." However, the arguments against this presumption are explained by Moula [101] as: "(1) he and many others fail to have a proper understanding of agriculture as an inherently technological practice which is different from 'nature'; (2) the notions of control and manipulation which are central to the concept of hubris are difficult to understand and use in the context of agriculture; (3) trying to establish a prima facie reason against GM crops runs into serious difficulty since many GM crops are profoundly different from each other; (4) even if we accept Sandler's argument of hubris, it actually plays no role when in our reasoning and evaluation of the moral status of different GM crops." So, rather than beginning with a presumption against the use of genetic modification, the approach should follow the advice given by the Nuffield Council on Bioethics (1999 and 2003) [77–78] that "policy towards GM crops should rest on a case-by-case analysis, with no general presumption in favor or against. Instead of general assertions, what was required was a sober assessment of the benefits and risks of particular applications against the feasible alternatives" [79]. Using the Markkula Centre's six lenses [2, 23] the debate between Sandler and Moula, while one of understanding that agriculture is an inherently technological practice and so is different from "nature," may still create a tension with some indigenous people's worldview, which sees humans as a part of nature.

3.2 Potential harm to human health

Uncertainty and skepticism about the safety of food from genetically modified plants to human health, which may include increased allergenicity, is a common concern [102–106]. This has been, and possibly still is, motivated by a lack of knowledge about the technology and the effects these foods can have on health [83, 107]. However, peer-reviewed studies have confirmed that food derived from genetically modified plants is as safe for consumption as its non-genetically modified counterparts [108–110]. After 15 years of consumption by hundreds of millions of people of foods derived from genetically modified crops, there were "no reported ill effects (or legal cases related to human health), despite many of the consumers coming from that most litigious of countries, the USA" [108]. In model animal studies, it was concluded that "oral consumption of a GM plant expressing a known allergen does not necessarily predispose to allergy but in fact may stimulate the development of a protective regulatory T-cell response and suppress the development of Allergic Airways Disease (AAD)" [111].

A panel of experts from diverse disciplines relevant to the safety of food from genetically modified crops for human consumption published a consensus statement that there was "no substantiated evidence of a difference in risks to human health between currently commercialized genetically engineered crops and conventionally bred crops" [112]. Further, more than 150 Nobel Prize winners have also stated that "scientific and regulatory agencies around the world have repeatedly and consistently found crops and foods improved through biotechnology to be as safe as, if not safer than, those derived from any other method of production. There has never been a single confirmed case of a negative health outcome for humans or animals from their consumption" [113]. However, providing this messaging, even from panels of experts, is often not sufficient to change ingrained public attitudes [114].

The demonstration that Brazil nut hypersensitivity in humans could result from the consumption of genetically modified soybeans expressing the Brazil nut 2S albumin protein, intended to improve its methionine content in seeds [115], highlighted the need to continue food safety testing as occurs in responsible regulatory systems. As a result, the soybean expressing the Brazil nut 2S albumin protein was not commercialized [116]. A recent summary of toxicology and health testing trials involving comparisons of GM crops and their non-GM equivalents, indicating either some concerns or little concern, demonstrates that biosafety testing procedures for genetically modified crops are effective [34]. Interestingly, non-genetically modified crops may also contain potentially toxic or allergenic compounds [117–119], but do not require any regulatory approvals for use.

The European Food Safety Authority has declared that the documents titled "Guidance for risk assessment of food and feed from genetically modified plants" and "Guidance on the environmental risk assessment of genetically modified plants" are "not relevant for the risk assessment of plants developed via SDN-1, SDN-2 or oligonucleotide-directed mutagenesis approaches if the genome of the final product does not contain exogenous DNA" [120]. In reality, SDN-1 and SDN-2 type gene edits are cisgenic (i.e., the genetic modification of a recipient plant with a natural gene from a crossable, sexually compatible plant) in nature [121, 122].

Biosafety reviews of new products derived from genetic modification and gene editing are considered an ethical response to determine the impacts on animal and human health when consumed [123, 124] and form the basis, along with environmental impact assessments, of gene technology regulatory systems worldwide [125]. Regulatory processes evaluating the safety of food derived from genetically modified plants "focus of risk assessment is primarily on identification of hazards (potential risks) and their characterization with respect to possible negative impacts on human/animal health and the environment, whereas potential beneficial impacts are usually not considered during risk assessment" [126]. An interesting comparison between Chile and Norway with regard to attitudes toward the use of genetic modification in the salmon industry highlights this issue. "In Norway, the anticipation of risks related to gene technology innovation is shaped by uncertainty about potential consequences. In Chile, by contrast, risk anticipation is approached with optimism about the opportunities gene technology may offer" [127]. As a result, the risks associated with this endeavor need to be addressed differently. In Norway, the focus should be on the voice of the scientific community in gene technology policy making, while in Chile, the emphasis should be on defining how to legitimize its development within the public.

3.3 The place of consumer satisfaction

The first commercial genetically modified crops focused on delivering input traits that primarily benefitted the farmer/producer through improved agricultural productivity but provided little benefit to the consumer [128]. As output traits [129, 130] are delivered that improve the nutritive value, taste, and storability of the food, and reduce environmental impacts (e.g., reduce greenhouse gas emissions), then consumers may be more accepting of the use of genetic modification. For horticultural plants, unique quality factors rather than yield per se must be the overriding breeding objective. Indeed, it has been stated that "a way must be found to incorporate consumer satisfaction in the selection process. Marketing attempts to obscure cultivar identification and change horticultural crops into commodities need to be resisted by the industry" [131].

3.4 Potential damage to the environment

Environmental risks that have been considered as potentially associated with genetically modified crops might include:

- Increased weediness – largely associated with the use of herbicide-resistant genetically modified crops and the resulting incorrect use of herbicides, causing herbicide resistance to emerge in weed species [132]. However, herbicide resistance is a global issue and not just restricted to the use of herbicide-tolerant genetically modified crops [133, 134].

- Secondary insect pests become dominant and may require other means of control [135].
- Development of resistant insect pest populations [136], which can be moderated using refuges [137].
- Negative impacts on beneficial insects, although evidence for this is scarce [138].
- Negative impacts on biodiversity [139], although this has been reviewed [34, 140] and the general conclusion is that the use of genetically modified crops has not significantly affected levels of genetic diversity within crop species.
- Potential ecological consequences, such as the potential for gene flow from genetically modified plants to wild relatives or the disruption of natural ecosystems [141], including genetic erosion of landraces [142], have been noted. An early review concluded that horizontal gene transfer from genetically modified plants poses negligible risks to human health or the environment [143]. Additionally, an extensive review of the coexistence of genetically modified and non-genetically modified plants has been recently published [144].

An extensive review by a panel of experts from diverse disciplines relevant to the environmental impacts of growing genetically modified crops, assembled under the auspices of the USA National Academies of Sciences, Engineering, and Medicine [112], concluded that there was "no conclusive evidence of cause-and-effect relationship between genetically engineered crops and environmental problems." However, it added that "the complex nature of assessing long-term environmental changes often made it difficult to reach definitive conclusions."

3.5 Negative impact on traditional farming practices

Genetic modification of crops has been condemned on the basis that it will result in negative social and economic impacts, particularly in developing countries among small farm holders, while possibly benefitting large farming enterprises in more industrialized countries [95]. There is little evidence for this, and in fact, some analyses indicate that the benefits of genetically modified crops are considerable for small farm holders [41]. Another consideration is that some African countries in the past have taken advice to adopt the European regulatory approach and removed the option for the use of genetically modified food and crops from their economies, despite the fact that in Africa, two-thirds of farmers are poor and in desperate need of new technologies to boost crop productivity [145]. Reliance on the European regulatory approach is, however, apparently changing [146]. Despite this, African farmers may still be disadvantaged by the availability and accessibility of seeds for genetically modified crops, their price, and the lack of adequate information [147].

3.6 Equitable access and/or unfair equitable access

The proliferation of intellectual property rights that relate to biotechnology, genes, and plants has been dubbed an "anti-commons" tragedy [148]. The argument has two levels – (1) intellectual property rights are primarily used as a trade instrument rather than as a means of supporting innovation, and (2) traditional farming communities viewed "seeds" as common property and therefore were freely shared. Some of these concerns are addressed in the International Treaty on Plant Genetic Resources for Food and Agriculture, which entered into force in 2004 [149]. An examination of possible conflicts between this International Treaty and other treaties, such as the Agreement on Trade-Related Aspects of Intellectual Property Rights (TRIPS), the Acts of the Union for the Protection of New Varieties of Plants (UPOV), and the Convention on Biological Diversity (CBD), which deal with similar subject matter but from quite different perspectives, indicated no apparent conflicts or legal concerns [149].

While some consider that patents and plant variety rights are "killing freedom to operate and crushing science with rules," for plant breeders and patent owners, strong IP protection offers an effective form of security, confirms that the product is unique, distinctive, and of value, and, importantly, brings returns on investment through licences and commercialization arrangements for further investment in the development of new traits and cultivars [150]. While there is no simple answer to whether intellectual property protection stimulates or constrains innovation and technology transfer, it is here to stay. Ideally, both individuals and organizations who use intellectual property protection need to do so while delivering maximum benefit to the end-user farmer. This is of particular importance when considering small-holder farmers in developing countries [96], such that the benefits of genetically modified crops are accessible to small-scale farmers and not limited to large corporations [59].

3.7 Food sovereignty

There is a view that "the expansive use of genetically modified organisms in agriculture is one of the most potent symbols of the worldwide threat to food justice and food sovereignty." [151]. The issue here is the dominating influence of multinational companies, who are accused of "stealing nature's harvest of diverse species, either by deliberately destroying biodiversity or by unintended biological pollution of species and ecosystems" [152]. The argument is that the tremendous diversity of crop cultivars, which have been developed by farmers over thousands of years, is now "under threat from genetic erosion and genetic piracy."

A perspective that has been used to justify the use of genetic modification of crops is the moral imperative for reducing poverty and hunger and promoting equity in the world's poorest countries [8, 153, 154]. However, others have argued that the principal driver for genetic modification was more commercial and self-centered and less altruistic and utilitarian [95], hence the resulting ethical dilemma. This relates back to which ethical lens is being used, and when this differs between individuals and groups, it can lead to different conclusions about what is right or wrong regarding genetically modified crops. Indeed, it has been recognized that in the world of biotechnology, and in particular genetic modification, the private sector (often associated with large multinational companies) has been both the developer of the technology (protected through patents) and the delivery vehicle to the end-user farmer [155]. From this, it is construed that "the private sector exploits opportunities which do not lead to the production of public goods or address the special needs of local ecosystems" [8]. There are, however, some exceptions to this, as witnessed by Wageningen University & Research (WUR), which will provide potential partners with free licenses to work on its patented CRISPR technology, so long as the license is applied to gene-editing of plants for non-profit applications [156]. This is a good example of where ethical tensions are recognized and institutional arrangements developed to address them. So, both technical innovation and institutional innovation are required to resolve the ethical tensions generated by genetic modification research. Institutional innovation is the deliberate creation or adaptation of rules, norms, and organizational structures that enable fair, effective, and inclusive functioning of technologies, often to support the adoption of new technologies and respond to different socio-ecological conditions [157, 158].

3.8 Excessive corporate dominance

The global food system and markets are dominated by a relatively small number of large multinational corporate companies positioned in the seed and agrochemical industries [159]. This consolidation is driven by financial incentives, technological change and complementarities, and a changing regulatory environment, resulting in an increased capacity to influence not only prices in the sector but also working conditions and compensation for food system workers, as well as consumer choice [160]. In addition, they can influence regulatory decisions, industry standards-setting, and the shaping of public discourse about the products which they sell. However, it has been argued that the seed sector has only modest economic weight and that the major influencers of the global food system are the downstream food retail sector [161, 162] and the agricultural and trade policies of the world's largest countries [163]. There is, however, at the very least, a widely held perception that multinational companies are dominant influencers of the global food systems [164], particularly with regard to oversight and incorporation of genetically modified crops [165, 166]. In some jurisdictions, such as the USA, only large multinationals can afford the cost and time associated with deregulating a genetically modified crop [167].

Plant breeding has evolved from being a largely academic pursuit in the early 20th century to one driven by significant commercial interests [24, 168]. An early concern regarding the use of genetic modification related to the traits being targeted and the fact that the work was primarily conducted by chemical companies, focusing on chemically dependent crop cultivars [169, 170] such as those providing herbicide resistance or pest resistance. It has been argued that "crop breeding and genetic modification are continuous with each other in many important ways but represent a momentous break from domestication because they move plant evolution off of farms and into centralized institutions" [171]. They argue that presenting genetic modification as a continuation of domestication "diminishes public understanding of these important processes and obscures the effects of industrial agriculture on in situ biodiversity and the practice of farming." It is deemed a ploy "by representatives of the biotechnology industry to silence meaningful debate on genetic modification." Furthermore, it fails to address concerns held by some in society regarding the potential social equity and environmental impacts of genetically modified plants [14].

One of the perceived, and possibly real, drivers of this change, from public good to commercial involvement, was the advent of intellectual property protection through Plant Variety Rights, trademarking, and patenting [172, 173]. This became more apparent in the latter third of the 20th century and saw the concomitant gradual move from government-funded plant breeding programmes in research institutions to ones managed and funded by commercial companies [168, 174]. Intellectual property protection resulted in the creation of proprietary products and ensured the plant breeder gained a return on their plant breeding investment through a royalty on sales of seed or, in some cases, as an end-point royalty on harvested grain, as occurs in Australia [175, 176]. However, during that transition, new genetic tools and technologies have greatly impacted the rate of genetic change. This has ensured that, while the world population has grown dramatically and, coincidentally, the proportion of the population in developed countries involved in agriculture has decreased remarkably [24], the ability to feed more people has become more efficient through not just increased use of fossil energy reserves but also through the breeding of improved cultivars [177–179].

3.9 Transparency and trust

To address any ethical concerns, it is essential to ensure transparency, open dialogue and public engagement in discussions involving plant biotechnology [59]. Science and technology, in general, are viewed by some as the cause of most of the problems facing society and humanity today, resulting in the move toward alternative lifestyles [95]. However, because of the commercial and profit motives of companies developing and marketing genetically modified crops, [103] observed that "consumer organizations, environmental groups, and scientists are considered to be more trustworthy than the biotech industry and government" [104, 180]. Building trust and fostering a more balanced understanding of genetic modification of crops requires effective communication, increased transparency in regulatory processes, and inclusive engagement with the public [181, 182]. To achieve this, recognition of different ethical perspectives and values will be needed to identify specific tensions and explore, in multi-stakeholder processes, how these can be addressed [5, 14].

4. Objectivity versus subjectivity in ethical decision-making

The argument to separate ethics from risk has been made [73, 183] but also criticized [184] because "the information necessary to assess risk from a scientific perspective is always the same; the ethical implications become apparent when the information has to be weighed in order to make a decision; that is, when responsibility, accountability, and justification have to be apportioned [73, 185]" [186]. By separating "risk" and "ethics" and dealing with risk regulation through applying objective science [187], it is argued that broader public debate is subsequently marginalized [184]. This would be a significant driver for the legislation used in Norway [127], which states that, as well as being safe for humans and the environment, "the production and use of GMOs should be ethically and socially justifiable in accordance with the principle of sustainable development" [188]. Ethically and socially justifiable principles of sustainable development would aim to balance the needs of the present with those of future generations, ensuring a harmonious coexistence between economic growth, environmental protection, and social equity [189–191]. So, for many, the concept of good ethics is more than just good reasoning based on evidence but also involves one's beliefs and values [192].

Distinguishing between genuine ethical concerns and prejudices will require an understanding of the nature and basis of the concerns raised. Is it acceptable that, for an ethical concern to be genuine, it must be based on fact/evidence and rational discourse, otherwise it could simply be labeled as prejudice resulting from misinformation and/or emotion? Then how does one manage ethical concerns resulting from cultural and religious beliefs? It is most often the case that the "intrinsic" ethical concerns about genetic modification being unnatural or the result of humankind "playing God" are manifested through philosophical or religious beliefs rather than scientific evidence [73; 193]. This is a challenging debate that is not easily resolved by consequentialist (outcomes) or procedural (process) ethics, since this is a debate about our relationship with nature. This highlights a fundamental ethical divide:

- Technocentric worldview, which views nature as modifiable for human benefit.
- Eco-centric or spiritual worldview, which views nature as sacred, interconnected, and not fully knowable or controllable.

This is especially relevant in indigenous or culturally grounded societies, where ethical reasoning includes ancestral relationships, spiritual values, and non-human agency. Hence, it is not resolvable by scientific evidence alone.

Moral theories, which "reflect the effort to think with clarity about very complicated areas of human thought, action, and conviction" should not be confused with the truth [3]. There is also a view that "an action is right, simply, because it is what duty or obligation requires," but that then begs the question of "what is our duty?" Another view is that "an action is right because of the consequences to which it leads" which in a utilitarian society results in "the greatest happiness for the greatest number of people," which then begs the question of "can we predict, with certainty, the consequences of our acts?" This view also "seems to disregard the rights and interests of individuals" and may go so far as to "sacrifice individuals to achieve the good for the group and demonstrates the difference of using utilitarian lens compared to a rights-based lens, as proposed by the Markkula Centre [23]. Then there is the view that "the reference point is not the collective good, but a specific good defined or identified by the theory of human nature in question," where "good is defined in terms of what is best in us, or normal or fully mature, or consistent with what God wants of us, which compares a common good lens compared to a virtue based lens in decision making. And while the scientific process has allowed us to discover and understand many things and as result achieve much, are we therefore "likely to conclude that because we can do something that we automatically should do it?" This logic is motivated by living in a society driven to solving problems but perhaps should be modified by understanding "that an additional argument is required to support any conclusion concerning what we should do" [3].

5. Institutional innovation to respond to ethical concerns

Institutional innovation refers to the deliberate creation or adaptation of rules, norms, and organizational arrangements that enable more inclusive, adaptive, and resilient plant biotechnology development, access, deployment, and distribution. It goes beyond technical innovation to address governance, coordination, and benefit-sharing mechanisms, ensuring that innovations in plant biotechnology are effectively adopted and socially legitimate [157, 158, 194].

For example, in Southwest China, institutional innovation supported participatory plant breeding by introducing Access and Benefit Sharing (ABS) contracts and market-based geographical indications, enabling farmers to benefit from new maize varieties and bridging gaps between local practices and national legislation [157, 158]. Similarly, pluralistic seed governance platforms integrate formal and farmer-led systems to ensure breeding outputs meet diverse agroecological and cultural needs [194]. Applied to genetically modified crops, institutional innovation could involve:

- A shift from process-based genetic modification regulations to trait-based risk assessment frameworks that evaluate ecological and health impacts rather than breeding methods [195].
- Multi-stakeholder governance forums, where farmers, scientists, consumer groups, and policymakers co-design genetic modification policies [17, 158],
- Public-interest intellectual property models, such as open-access or pooled IP [156], aim to foster trust and equitable access to genetically engineered traits.

6. Concluding comments

Balancing ethical requirements with the potential benefits of a technology will continue to be a challenge. However, perhaps optimistically, Weale [79], in his review on ethical arguments relevant to the use of GM crops, concluded that "we should avoid the fallacy of thinking that an ethical assessment will indicate a brake on technological development, particularly in cases where the technology addresses urgent human needs." Indeed, there are often different ethical priorities associated with plant genetic modification and gene editing. The ethical lenses of common good and utilitarianism accept the benefits of genetic modification, resulting in a potential to increase food supplies to alleviate hunger, reduce environmental impacts, and improve food quality and resulting human health benefits [153, 196, 197].

Ethical concerns about the use of biotechnology in agricultural systems must take into consideration the contrast between what has been referred to as traditional agriculture and modern agriculture. "Traditional agriculture applies an ecosystem approach to ensure harvest security and sustainability," where "productivity is often, though not always, relatively low" [6]. In modern agriculture, the use of genetic improvement and the advent of chemical fertilizers, chemical pest and disease control, irrigation, and mechanization meant that "the environment could increasingly be adapted to the requirements of individual crops," resulting in improved yields with more uniform products and security of supply [6]. Indeed, this agricultural revolution has enabled an increasing world population to be fed with fewer farmers on less land area because of markedly increased crop yields [154]. However, the downside of that advance has been the negative environmental impact on land, water, and the atmosphere. This has resulted in soil erosion and loss, decarbonization of soils through over-cultivation, increased weed issues through increased herbicide resistance, pollution of waterways with chemicals and animal waste, increased levels of carbon dioxide in the atmosphere (which has been associated with climate change), and now a possible plateau for some crop and pasture yields in some regions [198–203].

This, therefore, begs the question of where humankind goes now with agricultural systems that are under stress to supply the food required by a world where the population is still increasing, without further negative impacts on the environment? Returning to so-called traditional farming methods would most likely result in widespread hunger and starvation, even with improved distribution systems. Providing plants and possibly animals that lessen the impact of farming on the environment by including traits that reduce disease and pest impacts, lower greenhouse gas emissions, decrease the use of synthetic (and often mammalian-toxic) chemicals to control pests and disease, are more efficient in their use of nutrients and water, have improved nutritional and health characteristics, and save further land from conversion to agriculture – thus conserving plant biodiversity in natural habitats, must be an option worth considering within any ethical context. The primary objective of plant breeding is to develop crops that better suit desired requirements, and this is achieved through exploiting genetic diversity. Genetic modification and gene editing are means to extend that genetic diversity for the ultimate good of humankind. Regulations that are cost-effective, realistic, and provide confidence to consumers are still required to monitor biosafety/biorisk for human health and the environment. However, in the end, it is the consumer who will decide whether biotechnology and genetic modification are acceptable approaches for future crop improvement. A quote from 30 years ago is still very pertinent – "The level of trust that has characterised science and its relationship to society will endure only if the scientific community devotes itself to exemplifying and transmitting the values associated with ethical scientific conduct" [204]. To achieve this outcome, all aspects of scientific endeavor, including biotechnology, must impose self-scrutiny through free and open debate and peer review [205]. Scientists using genetic modification need to reflect on the ethical lens they use and how these may be in tension with the ethical lens of other groups they may engage with, such that different values need to be included in the debate about genetically modified crops. This is the anticipation, reflexivity, and inclusion dimensions of Responsible Innovation [76].

Ethical debate on the use and value of foods derived from genetically modified plants is unlikely to result in simple answers and universal agreement and will involve both technical and institutional innovation if progress on the ethical differences is to be made [157, 158]. A sensible approach is to engage with and respect different perspectives, base any discussion on scientific evidence, and ensure that the discussion is centered on understanding the impacts on human well-being, environmental impacts, and social justice. To achieve this, ethical frameworks and processes, such as the framework for ethical decision-making [23], can be used to guide discussion and respond to ethical concerns about plant breeding, plant biotechnology, and genetic modification.

The ethical concerns societal groups have toward plant breeding tend to focus on justice, rights, and care ethics (as well as procedural ethics, e.g., the right to be consulted, participate, and be informed), while research on the benefits and risks of genetic modification tends to prioritize utilitarian, common good, and virtue ethics (as well as consequential ethics, i.e., whether the outcomes of genetic modification are good and beneficial). This leads to science responding to societal ethical concerns from a different ethical position, exacerbating tensions rather than resolving them. A more widespread use of the Markkula Center's six lenses by the research community may assist with some resolution. They are a potentially fruitful "ethical tool" to assist in resolving normative conflicts by:

- Clarifying the values at stake (e.g., rights vs. common good).
- Providing structured approaches to evaluate competing ethical concepts of what is right.
- Encouraging balanced decisions that consider consequences, fairness, character, and relationships.

Additionally, institutional innovation must accompany technical innovation to resolve these normative conflicts, which will mean changing the rules, structures, and ways that organizations, governments, or communities work together [157, 158]. This may then allow problems and tensions to be solved more effectively, fairly, or sustainably. It isn't just about new technologies or products, but about rethinking how decisions are made, who gets to participate, and how the development and deployment of plant biotechnologies adapt to new challenges.

However, there is a fundamental ethical tension that may not be resolvable with some groups, especially indigenous communities, concerning the debate about our relationship with nature and the tension between a technocentric worldview and an eco-centric or spiritual worldview.

Conflict of Interest

JRC is employed by Grasslanz Technology Ltd, which has an R&D investment portfolio that includes both genetic modification and gene editing of forages and microbes to provide mitigating solutions to current environmental and animal welfare issues facing both New Zealand and other pastoral economies.

Author details

John R. Caradus[1*] and James A. Turner[2]

1 Grasslanz Technology Ltd., Hamilton, New Zealand

2 New Zealand Institute for Bioeconomy Science Ltd., Hamilton, New Zealand

*Address all correspondence to: john.caradus@grasslanz.com

References

[1] Britannica 2025. Ethics. Available from: https://www.britannica.com/topic/ethics-philosophy [Accessed: 2025-September-28]

[2] Center M. A framework for ethical decision making. Santa Clara Markkula Center for applied ethics. 2021. Available from: https://www.scu.edu/ethics/ethics-resources/a-framework-for-ethical-decision-making/ [Accessed: 2025-September-29]

[3] Sheldon M Fundamentals of ethics. In: Biotechnology and ethics: A blueprint for the future. 1996. Available from: https://www.iatp.org/sites/default/files/Fundamentals_of_Ethics.htm [Accessed: 2025-October-29]

[4] Netherlands Commission on Genetic Modification. Towards an integrated framework for the assessment of social and ethical issues in modern biotechnology. COGEM advisory report to the state secretary for housing, planning and the environment (CGM/030618-02). Pp.62. 2003. Available from: https://cogem.net/app/uploads/2019/07/030618-02en1.pdf [Accessed: 2025-October-03]

[5] Veflen N, Bánáti D. Chapter 54 - Ethics in food safety management. In Andersen V, Lelieveld H, Motarjemi Y, editors. Food Safety Management. Second ed. San Diego: Academic Press; 2023. p. 1081–1088

[6] Hardon JJ Ethical issues in plant breeding, biotechnology and conservation. In: Ethics and equity in conservation and use of genetic resources for sustainable food security. Proceedings of a workshop to develop guidelines for the CGIAR, Foz do Iguacu, Brazil, IPGRI, 21-25 April 1997. p. 43–50. Available from: https://hdl.handle.net/10947/498

[7] Straughan R Analysis of ethical principles arising from the Workshop. In: Ethics and equity in conservation and use of genetic resources for sustainable food security. Proceedings of a workshop to develop guidelines for the CGIAR, 21-25 April 1997, Foz do Iguacu, Brazil, IPGRI. p. 173–180. Available from: https://hdl.handle.net/10947/498

[8] Serageldin I Equity and ethics: Twin challenges, twin opportunities. In: Ethics and equity in conservation and use of genetic resources for sustainable food security. Proceedings of a workshop to develop guidelines for the CGIAR, 21-25 April 1997, Foz do Iguacu, Brazil, IPGRI. Pp. 1–6. Available from: https://hdl.handle.net/10947/498

[9] Weinbaum C, Landree E, Blumenthal MS, Piquado T, Gutierrez Gaviria CI. Ethics in Scientific Research: An Examination of Ethical Principles and Emerging Topics. RAND Corporation: Santa Monica, CA; 2019. p. 118. Available from: https://apps.dtic.mil/sti/html/tr/AD1096839/ [Accessed: 2025-November-30]

[10] Fossey A. Bioethics in Agricultural Research and Research Management. In Loebenstein G, Thottappilly G, editors. Agricultural Research Management. 1st ed. Dordrecht, Netherlands: Springer; 2007. p. 121–147. DOI: 10.1007/978-1-4020-6057-1_7

[11] Potter VR. Bioethics: The science of survival. Perspectives in Biology and

Medicine. 1970;**14**:127–153. https://muse.jhu.edu/pub/1/article/405198/pdf

[12] Macer DRJ Bioethics is love of life: An alternative textbook. Christchurch, N.Z.: Eubios Ethics Institute. 1998. Available from: https://eubios.info/BLL.htm [Accessed: 2025-September-06]

[13] Moula P, Sandin P. Evaluating ethical tools. Metaphilosophy. 2015;**46**:263–279. DOI: 10.1111/meta.12130

[14] Noll S. A framework for thawing value conflicts in the GMO debate. In Vallor S, editor. The Oxford Handbook of Philosophy of. UK: Technology Oxford University Press, 2020:496–515. DOI: 10.1093/oxfordhb/9780190851187.013.30

[15] Hayes A Understanding codes of ethics: Types and their practical uses. 2025. Available from: https://www.investopedia.com/terms/c/code-of-ethics.asp [Accessed: 2025-October-04]

[16] McCarthy J The five fundamental principles of ethical behaviour. 2020. Available from: https://www.jmcc.ie/the-five-fundamental-principles-of-ethical-behaviour/ [Accessed: 2025-October-04]

[17] Pimbert MP, Barr B. Let the people decide: Citizen deliberation on the role of GMOs in Mali's agriculture. Agriculture and Human Values. 2021;**38**:1097–1122. DOI: 10.1007/s10460-021-10221-1

[18] FoodSafety Institute. Food Biotechnology. Ethical Considerations Surrounding GM Foods. 2024. Available from: https://foodsafety.institute/food-biotechnology/ethical-considerations-gm-foods/ [Accessed: 2025-October-04]

[19] Mepham B, Kaiser M, Thorstensen E, Tomkins S, Millar K Ethical Matrix User Manual, The Netherlands: Agricultural Economics Research Institute. 2006. Available from: https://edepot.wur.nl/216589 [Accessed: 2025-November-30]

[20] Kaiser M, Millar K, Thorstensen E, Tomkins S. Developing the ethical matrix as a decision support framework: GM fish as a case study. Journal of Agricultural and Environmental Ethics. 2007;**20**:65–80. DOI: 10.1007/s10806-006-9023-8

[21] Forsberg EM. The ethical matrix - A tool for ethical assessments of biotechnology. Global Bioethics. 2004;**17**:167–172. DOI: 10.1080/11287462.2004.10800856

[22] Beekman V, Bakker E, Baranzke H, Baune O, Deblonde M, Forsberg EM, de Graaff R, Ingensiep H-W, Lassen J, Mepham B, Nielsen AP, Tomkins S, Thorstensen E, Millar K, Skorupinski B, Brom F, Kaiser M, Sandoe P. 2006. "Ethical bio-technology assessment tools for agriculture and food production." Final Report Ethical Bio-TA Tools. (QLG6-CT-2002-02594). Available from: https://estframe.net/ethical_bio_ta_tools_project/content_2/text_2c81d261-b7a8-43e8-8f1e-d724b43e2ba3/1346076907647/et1_final_report_binnenwerk_59p.pdf

[23] Center M. A framework for ethical decision making. santa clara markkula center for applied ethics. 2025. Available from: https://www.scu.edu/media/ethics-center/ethical-decision-making/A-Framework-for-Ethical-Decision-Making-2025.pdf [Accessed: 2025-September-29]

[24] Orton TJ. Horticultural Plant Breeding. London, UK: Academic Press,

Elsevier Science; 2019. p. 410p. ISBN:9780128155707,0128155701. https://www.google.co.nz/books/edition/Horticultural_Plant_Breeding/rte_DwAAQBAJ?hl=en&gbpv=1

[25] Ahmad F. Historical perspectives of cutting-edge genome editing techniques. In Al-Khayri JM, Sattar MN, Sopory SK, Jain SM, editors. Genome Editing for Crop Improvement: Theory and Methodology. Wallingford, UK: CABI Books, CABI, 2025:1–16. DOI: 10.1079/9781800622517.0001

[26] Duvick DN. Plant breeding, an evolutionary concept. Crop Science. 1996;**36**:539–548. DOI: 10.2135/cropsci1996.0011183X00 3600030001x

[27] Duvick DN, Johnson P, Falk DE, Elliot T. The current state of plant breeding: How did we get hereIn Sligh M, Lauffer L, editors. Summit Proceedings: Summit on Seeds and Breeds for 21st Century Agriculture. Pittsboro, North Carolina, USA: Rural Advancement Foundation International; 2004. p. 71–92. Available from: https://www.leopold.iastate.edu/files/pubs-and-papers/2003-09-summit-proceedings-seeds-breeds-21st-century-agriculture.pdf#page=81 [Accessed: 2025-October-04]

[28] Hallauer AR. Evolution of plant breeding. Crop Breeding and Applied Biotechnology. 2011;**11**:197–206. DOI: 10.1590/S1984-70332011000300001

[29] Janick J. The origins of horticultural technology and science. XXVII International Horticultural Congress-IHC2006: Global Horticulture: Diversity and Harmony, an Introduction to IHC2006. 2006;**759**:41–60

[30] Miller HI, Morandini P. Plant domestication, the brave old world of genetic modification. Advances in Botanical Research. 2018;**86**:1–15. DOI: 10.1016/BS.ABR.2017.11.001

[31] Osei M, Asante M, Agyeman A, Adebayo M, Adu-Dapaah H. Plant Breeding: A tool for achieving food sufficiency. In Nandwani D, editor. Sustainable Horticultural Systems. Sustainable Development and Biodiversity. Vol. **2**. Cham, Heidelberg: Springer; 2014. p. 253–274. DOI: 10.1007/978-3-319-06904-3_11

[32] Schlegel RHJ. History of Plant Breeding. Boca Raton, USA: CRC Press, Taylor Francis Group; 2017. p. 330p. ISBN:9781351588966,1351588966

[33] Bonciu E, Iancu P, Bonea D. The genetically modified plants - acceptance and controversies. Journal of Horticulture, Forestry and Biotechnology. 2013;**17**:166–172

[34] Caradus JR. Intended and unintended consequences of genetically modified crops – Myth, fact and/or manageable outcomes? New Zealand Journal of Agricultural Research. 2022;**66**:519–619. DOI: 10.1080/00288233.2022.2141273

[35] Ghimire BK, Yu CY, Kim W-R, Moon H-S, Lee J, Kim SH, Chung IM. assessment of benefits and risk of genetically modified plants and products: Current Controversies and Perspective. Sustainability. 2023;**15**:1722. DOI: 10.3390/su15021722

[36] Caradus JR. Perceptions of plant breeding methods – From 'phenotypic selection' to 'genetic modification' and 'new breeding technologies'. New Zealand Journal of Agricultural

Research. 2024;**67**:621–669. DOI: 10.1080/00288233.2023.2187425

[37] Gatew H, Mengistu K. Genetically modified foods (GMOs); a review of genetic engineering. Journal of Life Science and Biomedicine. 2019;**9**:157–163. DOI: 10.36380/scil.2019.jlsb25

[38] Gepts P. A comparison between crop domestication, classical plant breeding, and genetic engineering. Crop Science. 2002;**42**:1780–1790. DOI: 10.2135/cropsci2002.1780

[39] Jauhar PP. Modern biotechnology as an integral supplement to conventional plant breeding: The prospects and challenges. Crop Science. 2006;**46**:1841–1859. DOI: 10.2135/cropsci2005.07-0223

[40] Ricroch A. The evolution of agriculture and tools for plant innovation. In Ricroch A, Chopra S, Kuntz M, editors. Plant Biotechnology. Cham, Heidelberg: Springer, 2021:3–15. DOI: 10.1007/978-3-030-68345-0_1

[41] Klümper W, Qaim M. A meta-analysis of the impacts of genetically modified crops. PLoS One. 2014;**9**:e111629. DOI: 10.1371/journal.pone.0111629

[42] Abebe A, Tafa Z. A review on: Impact of genetic engineering in biotic stresses resistance crop breeding. International Journal for Research in Applied Sciences and Biotechnology. 2022;**9**:243–250. DOI: 10.31033/ijrasb.9.2.18

[43] Khokhar A, Shahbaz M, Maqsood MF, Zulfiqar U, Naz N, Iqbal UZ, Sara M, Aqeel M, Khalid N, Noman A, Zulfiqar F, Al Syaad KM, AlShaqhaa MA. Genetic modification strategies for enhancing plant resilience to abiotic stresses in the context of climate change. Functional and Integrative Genomics. 2023;**23**: Article283. DOI: 10.1007/s10142-023-01202-0

[44] Kumar K, Gambhir G, Dass A, AK AKT, Singh A, Kumar Jha A, Yadava P, Choudhary M, Rakshit S. Genetically modified crops: Current status and future prospects. Planta. 2020;**251**:91. DOI: 10.1007/s00425-020-03372-8

[45] Tohidfar M, Khosravi S. Transgenic crops with an improved resistance to biotic stresses. A review. Biotechnology, Agronomy, Society and Environment. 2015;**19**:62–70. Available from: https://popups.uliege.be/1780-4507/index.php?id=11844

[46] Ahmad P, Ashraf M, Younis M, Hu X, Kumar A, Akram NA, Al-Qurainy F. Role of transgenic plants in agriculture and biopharming. Biotechnology Advances. 2012;**30**: 524–540. DOI: 10.1016/j.biotechadv.2011.09.006

[47] Khalil AM. The genome editing revolution: Review. Journal of Genetic Engineering and Biotechnology. 2020;**18**:1–16. DOI: 10.1186/S43141-020-00078-Y

[48] Graff GD, Hochman G, Zilberman D. The political economy of agricultural biotechnology policies. AgBioForum. 2009;**12**:34–46. Available from: https://hdl.handle.net/10355/1418

[49] Smith V, Wesseler JHH, Zilberman D. New plant breeding technologies: An Assessment of the political economy of the regulatory environment and implications for sustainability. Sustainability. 2021;**13**:3687. DOI: 10.3390/su13073687

[50] Wesseler J. Biotechnologies and agrifood strategies: Opportunities, threats and economic implications. Bio-based and Applied Economics. 2014;**3**:187–204. DOI: 10.22004/AG.ECON.196653

[51] Jorasch P. Potential, Challenges, and threats for the application of new breeding techniques by the private plant breeding sector in the EU. Frontiers in Plant Science. 2020;**11**:582011. DOI: 10.3389/fpls.2020.582011

[52] Zilberman D, Kaplan S, Kim E, Hochman G, Graff G. Continents divided: Understanding differences between Europe and North America in acceptance of GM crops. GM Crops and Food. 2013;**4**:202–208. DOI: 10.4161/gmcr.26981

[53] Chandler SF, Sanchez C. Genetic modification; the development of transgenic ornamental plant varieties. Plant Biotechnology Journal. 2012;**10**:891–903. DOI: 10.1111/j.1467-7652.2012.00693.x

[54] Entine J, Felipe MSS, Groenewald J, Kershen DL, Lema MA, McHughen A, Nepomuceno AL, Ohsawa R, Ordonio RL, Parrott WA, Quemada H, Ramage C, Slamet-Loedin I, Smyth SJ, Wray-Cahen D. Regulatory approaches for genome edited agricultural plants in select countries and jurisdictions around the world. Transgenic Research. 2021;**30**:551–584. DOI: 10.1007/s11248-021-00257-8

[55] Gleim S, Lubieniechi S, Smyth SJ. CRISPR-Cas9 Application in Canadian public and private plant breeding. The CRISPR Journal. 2020;**3**:44–51. DOI: 10.1089/crispr.2019.0061

[56] Smyth SJ. Global status of the regulation of genome editing technologies. CABI Reviews. 2019;1–6. DOI: 10.1079/PAVSNNR201914021

[57] Jenkins D, Dobert RC, Atanassova A, Pavely C. Impacts of the regulatory environment for gene editing on delivering beneficial products. In Vitro Cellular & Developmental Biology – Plant. 2021;**57**:1–18. DOI: 10.1007/S11627-021-10201-4

[58] Brookes G, Smyth SJ. Risk-appropriate regulations for gene-editing technologies. GM Crops and Food. 2024;**15**:1–14. DOI: 10.1080/21645698.2023.2293510

[59] The Bloom Up. What is plant biotechnology? discovering its role in sustainable agriculture and beyond. 2023. Available from: https://www.thebloomup.com/what-is-plant-biotechnology/#:~:text=Plant%20biotechnology%20is%20a%20field%20that%20combines%20genetics%2C,and%20develop%20new%20plant%20varieties%20with%20desirable%20traits [Accessed: 2025-September-29]

[60] National institute of food and agriculture. Plant Biotechnology. 2025. Available from: https://www.nifa.usda.gov/grants/programs/biotechnology-programs/plant-biotechnology [Accessed: 2025-September-29]

[61] Caccamo M The gene editing revolution is only just beginning. Science for Sustainable Agriculture. 2025. Available from: https://www.scienceforsustainableagriculture.com/mariocaccamo8 [Accessed: 2025-October-02]

[62] Songstad DD, Petolino JF, Voytas DF, Reichert NA. Genome editing of plants. Critical Reviews in Plant Sciences. 2017;**36**:1–23. DOI: 10.1080/07352689.2017.1281663

[63] Bhagtaney L, Sundarrajan P. An overview of tools for genome editing: ZFNs, mega nucleases, and TALENs. In Pandita D, Pandita A, editors. CRISPR/Cas-mediated Genome Editing in Plants. New York, USA: Apple Academic Press, 2023:37–64. DOI: 10.1201/9781003331759

[64] Doudna JA, Charpentier E. The new frontier of genome engineering with CRISPR-Cas9. Science. 2014;**346**:1258096. DOI: 10.1126/science.1258096

[65] FAO, FAO. Food and Agriculture Organization of the United Nations. Rome: FAO. p. 86; 2022. DOI: 10.4060/cc3579en

[66] Menz J, Modrzejewski D, Hartung F, Wilhelm R, Sprink T. Genome edited crops touch the market: A view on the global development and regulatory environment. Frontiers in Plant Science. 2020;**11**:586027. DOI: 10.3389/fpls.2020.586027

[67] Prakash A, Mandal S, Ranjan S, Sood N, Bahadur Singh RB, Kumar S, Kumari N. Ethical, societal issues, and future challenges of plant CRISPR technologies. In Abd-Elsalam KA, Chen J-T, editors. Genome Modified Plants and Microbes in Food and Agriculture. Cambridge, Massachusetts, USA: CRISPR Technology for Combating Plant Pests and Pathogens, Academic Press; 2026. p. 281–307. DOI: 10.1016/B978-0-443-43920-9.00013-0

[68] Rodriguez E. Ethical issues in genome editing using crispr/Cas9 system. Journal of Clinical Research and Bioethics. 2016;**7**:1–4. DOI: 10.4172/2155-9627.1000266

[69] Zhang D, Hussain A, Manghwar H, Xie K, Xie S, Zhao S, Larkin RM, Qing P, Jin S, Ding F. Genome editing with the CRISPR-Cas system: An art, ethics and global regulatory perspective. Plant Biotechnology Journal. 2020;**18**:1651–1669. DOI: 10.1111/PBI.13383

[70] Pirscher F, Theesfeld I The ethical dilemma with governing CRISPR/CAS genome editing. In: Springer S, Grimm H, editors. "Professionals in Food Chains". Proceedings of the EurSafe Conference. 13-16 June 2018. Vienna, Austria. Wageningen Academic Publishers, p. 419–423. https://papers.ssrn.com/sol3/papers.cfm?abstract_id=3226805

[71] Diptaningsari D, Martono E. Considering ethics in agricultural experimental research with special mention to agricultural biotechnology research. Journal Perlindungan Tanaman Indonesia. 2021;**25**:10–20. DOI: 10.22146/JPTI.57790

[72] Elisabetsky E. Sociopolitical, economical and ethical issues in medicinal plant research. Journal of Ethnopharmacology. 1991;**32**:235–239. DOI: 10.1016/0378-8741(91)90124-V

[73] Straughan R. Ethics, morality and crop biotechnology. 1. Intrinsic concerns. Outlook on Agriculture. 1995;**24**:187–192. DOI: 10.1177/003072709502400309

[74] Sinhmar D, Purohit R. Patenting life: Global perspectives on intellectual property in genetic and biotechnological innovation amidst ethical and legal challenges. Journal on Development of Intellectual Property and Research. 2025;**1**:29–40. DOI: 10.64618/z3gqsk52

[75] Schomberg R Towards responsible research and innovation in the information and communication technologies and security technologies fields. LU: Publications Office of the European Union. 224p.2011. Available from: https://data.europa.eu/doi/10.2777/58723 [Accessed: 2025-November-02]

[76] Stilgoe J, Owen R, Macnaghten P. Developing a framework for responsible innovation. In Maynard A, Stilgoe J, editors. The Ethics of Nanotechnology, Geoengineering, and Clean Energy. Routledge: London, UK; 2020. p. 347–359. https://www.taylorfrancis.com/chapters/edit/10.4324/9781003075028-22/developing-framework-responsible-innovation-jack-stilgoe-richard-owen-phil-macnaghten

[77] Nuffield Council on Bioethics. Genetically modified crops: The ethical and social issues. Nuffield Council on Bioethics. 1999. Available from: https://www.nuffieldbioethics.org/publication/genetically-modified-crops-the-ethical-and-social-issues/ [Accessed: 2025-September-29]

[78] Nuffield Council on Bioethics. The use of genetically modified crops in developing countries. Nuffield Council on Bioethics. 2003. Available from: https://www.nuffieldbioethics.org/publication/the-use-of-gm-crops-in-developing-countries/ [Accessed: 2025-September-29]

[79] Weale A. Ethical arguments relevant to the use of GM crops. New Biotechnology. 2010;**27**:582–587. DOI: 10.1016/j.nbt.2010.08.013

[80] Hendriks S, Giesbertz NAA, Bredenoord AL, Repping S. Reasons for being in favour of or against genome modification: A survey of the Dutch general public. Human Reproduction Open. 2018:hoy008. DOI: 10.1093/hropen/hoy008

[81] CGIAR. Ethics and equity in conservation and use of genetic resources for sustainable food security. Proceedings of a workshop to develop guidelines for the CGIAR, 21-25 April 1997, Foz do Iguacu, Brazil, IPGRI. Pp. 208. https://hdl.handle.net/10947/498

[82] Gregorowius D, Lindemann-Matthies P, Huppenbauer M. Ethical discourse on the use of genetically modified crops: A review of academic publications in the fields of ecology and environmental ethics. Journal of Agricultural and Environmental Ethics. 2012;**25**:265–293. DOI: 10.1007/s10806-011-9330-6

[83] Grunert KG, Lähteenmäki L, Nielsen NA, Poulsen GB, Ueland Ø, Åström A. Consumer perceptions of food products involving genetic modification: Results from a qualitative study in four Nordic countries. Food Quality and Preference. 2001;**12**:527–542. DOI: 10.1016/S0950-3293(01)00049-0

[84] Korthals M. Taking consumers seriously: Two concepts of consumer sovereignty. Journal of Agricultural and Environmental Ethics. 2001;**14**:201–215. DOI: 10.1023/A:1011356930245

[85] Meijboom FLB, Verweij MF, Brom FWA. You eat what you are: Moral dimensions of diets tailored to one's genes. Journal of Agricultural and Environmental Ethics. 2003;**16**:557–568. DOI: 10.1023/B:JAGE.0000004964.85970.8d

[86] Scott SE, Inbar Y, Wirz CD, Brossard D, Rozin P. An overview of attitudes toward genetically engineered food. Annual Review of Nutrition.

2018;**38**:459–479. DOI: 10.1146/annurev-nutr-071715-051223

[87] Vega Rodríguez A, Rodríguez-Oramas C, Sanjuán Velázquez E, Hardisson de la Torre A, Rubio Armendáriz C, Carrascosa IC. Myths and realities about genetically modified food: A risk-benefit analysis. Applied Sciences. 2022;**12**:2861. DOI: 10.3390/app12062861

[88] Bredahl L. Consumers cognitions with regard to genetically modified foods. Results of a qualitative study in four countries. Appetite. 1999;**33**:343–360. DOI: 10.1006/appe.1999.0267

[89] Roman S, Sanchez-Siles LM, Siegrist M. The importance of food naturalness for consumers: Results of a systematic review. Trends in Food Science and Technology. 2017;**67**:44–57. DOI: 10.1016/j.tifs.2017.06.01

[90] Rozin P. Naturalness judgments by lay Americans: Process dominates content in judgments of food or water acceptability and naturalness. Judgment and Decision Making. 2006;**1**:91–97. DOI: 10.1017/S1930297500002308

[91] Verhoog H, Matze M, van Bueren EL, Baars T. The role of the concept of the natural (naturalness) in organic farming. Journal of Agricultural and Environmental Ethics. 2003;**16**: 29–49. DOI: 10.1023/A:1021714632012

[92] Chambers E, Castro M. What is "natural"? Consumer responses to selected ingredients. Foods. 2018;**7**:65. DOI: 10.3390/foods7040065

[93] Rozin P, Fischler C, Shields-Argeles C. European and American perspectives on the meaning of natural. Appetite. 2012;**59**:448–455. DOI: 10.1016/j.appet.2012.06.001

[94] Grunert KG, Bredahl L, Scholderer J. Four questions on European consumers' attitudes toward the use of genetic modification in food production. Innovative Food Science and Emerging Technologies. 2003;**4**:435–445. DOI: 10.1016/S1466-8564(03)00035-3

[95] Jaffe WR Biotechnology and biosafety: Ethical issues and science policy. In: Ethics and equity in conservation and use of genetic resources for sustainable food security. Proceedings of a workshop to develop guidelines for the CGIAR, Foz do Iguacu, Brazil, IPGRI. 21-25 April 1997, p. 32–36. Available from: https://hdl.handle.net/10947/498

[96] FoodSafety Institute. Food biotechnology. Addressing ethical, safety, and environmental concerns in food biotechnology. 2023. Available from: https://foodsafety.institute/food-biotechnology/ethical-safety-environmental-food-biotechnology/ [Accessed: 2025-October-04]

[97] Bekele-Alemu A, Dessalegn-Hora O, Safawo-Jarso T, Ligaba-Osena A. Rethinking progress: Harmonizing the discourse on genetically modified crops. Frontiers in Plant Science. 2025;**16**:1547928. DOI: 10.3389/fpls.2025.1547928

[98] Oh J, Ezezika OC. To label or not to label: Balancing the risks, benefits and costs of mandatory labelling of GM food in Africa. Agriculture and Food Security. 2014;**3**:8. DOI: 10.1186/2048-7010-3-8

[99] Eldred JS. Analysis & perspective: Labelling GMO-derived food ingredients: A recipe for misinformation. World Food Regulation Review. 1999;**14**:16

[100] Sandler R. An aretaic objection to agricultural biotechnology. Journal of Agricultural and Environmental Ethics. 2004;**17**:301–317. DOI: 10.1023/B:JAGE.0000033078.05859.08

[101] Moula P. GM crops, the hubris argument and the nature of agriculture. Journal of Agricultural and Environmental Ethics. 2015;**28**:161–177. DOI: 10.1007/s10806-014-9526-7

[102] Bredahl L. Determinants of consumer attitudes and purchase intentions with regard to genetically modified food – results of a cross-national survey. Journal of Consumer Policy. 2001;**24**:23–61. DOI: 10.1023/A:1010950406128

[103] Costa-Font M, Gil JM, Traill WB. Consumer acceptance, valuation of and attitudes towards genetically modified food: Review and implications for food policy. Food Policy. 2008;**33**:99–111. DOI: 10.1016/j.foodpol.2007.07.002

[104] Gaskell G, Allum N, Stares SE, in B. Eurobarometer 58.0. A Report to the EC Directorate General for Research from the Project "life Sciences in European Society". QLG7-CT-1999-00286. European Commission: Brussels; 2003. p. 44. Available from: https://d1wqtxts1xzle7.cloudfront.net/66432478/Europeans_and_Biotechnology_in_200220210421-10916-570xvd.pdf?1738411174=&response-content-disposition=inline%3B+filename%3DEuropeans_and_Biotechnology_in_2002.pdf&Expires=1759634809&Signature=EjIj4Axgs4fB9HT2pknzo6983gvT0iRngqfd5F~mVHHPY1g9kvYTwvI3P52PQCHgZxP4zpGTQ9Sadd67Py6MAz5xtaEIXAulmyHWIy2rjvUteckSX1ZtOURI9Y9O49FKOGveylaarlKCcAIsFJYev8By4XGZdgCrbmX4AHZA0dam8wdVnbkoZdmBgBpTBkeO2fBmiQlNYEzM4hOf99D2E2jVz7WQH7Atp~NtGstxdrpe3IWAKAumUtWzGIIqJYo7KJJe5DngmT6XAqhAkSB5LQYoGHI9CdLMjeIqylQ0zakGTJSpVlE1jkjSNrzZSNId1ELaSIPAKmMZEjGwQIeqkw__&Key-Pair-Id=APKAJLOHF5GGSLRBV4ZA [Accessed: 2025-October-05]

[105] Martinez-Poveda A, Molla-Bauza MB, Gomis FJC, Martinez LMC. Consumer-perceived risk model for the introduction of genetically modified food in Spain. Food Policy. 2009;**34**:519–528. DOI: 10.1016/j.foodpol.2009.08.001

[106] Shaw A. It just goes against the grain. Public understandings of genetically modified (GM) food in the UK. Public Understanding of Science. 2002;**11**:273. DOI: 10.1088/0963-6625/11/3/305

[107] Frewer LI, Howard C, Sheperd R. Genetic engineering and food: What determines consumer acceptance? British Food Journal. 1995;**97**:31–36. DOI: 10.1108/00070709510100118

[108] Key S, Jk-c M, Drake PMW. Genetically modified plants and human health. Journal of the Royal Society of Medicine. 2008;**101**:290–298. DOI: 10.1258/jrsm.2008.070372

[109] Thomas JA. Safety of foods derived from genetically modified plants. Texas Medicine. 2003;**99**:66. https://www.ncbi.nlm.nih.gov/pubmed/12674981

[110] Tijjani MB, AM I, AT I. 2026. Genetically-modified crop risk assessment and biosafety: Integrative meta-analysis and communication strategy. In Ogwu M, and Izah S, editors. Mitigation Strategies and Risk Management for Food Contaminants. Hershey, Pennsylvania, USA: IGI Global

Scientific Publishing. p. 243–290. DOI: 10.4018/979-8-3373-7052-1.ch008

[111] Prescott AE, Hogan SP. Genetically modified plants and food hypersensitivity diseases: Usage and implications of experimental models for risk assessment. Pharmacology and Therapeutics. 2006;**111**:374–383. DOI: 10.1016/j.pharmthera.2005.10.005

[112] NASEM. National Academies of Science, Engineering and Medicine. Genetically Engineered Crops: Experiences and Prospects. Washington, DC, USA: National Academies Press; 2016. p. 606. DOI: 10.17226/23395

[113] Agre P, Alferov ZI, Altman S, Amano H, Ambros V, Arber W, Arnold FH, Axel R, Baltimore D, Barish BC, et al. Laureates letter supporting precision agriculture (GMOS). 2016. Available from: https://supportprecisionagriculture.org/nobel-laureate-gmo-letter_rjr.html [Accessed: 2025-October-04]

[114] Landrum AR, Hallman WK, Jamieson KH. Examining the impact of expert voices: Communicating the scientific consensus on genetically-modified organisms. Environmental Communication. 2019;**13**:51–70. DOI: 10.1080/17524032.2018.1502201

[115] Nordlee JA, Taylor SL, Townsend JA, Thomas LA, Bush RK. Identification of a Brazil-nut allergen in transgenic soybeans. New England Journal of Medicine. 1996;**334**:688–692. DOI: 10.1056/NEJM199603143341103

[116] Herman EM. Genetically modified soybeans and food allergies. Journal of Experimental Botany. 2003;**54**:1317–1319. DOI: 10.1093/jxb/erg164

[117] Halford NG, Shewry PR. Genetically modified crops: Methodology, benefits, regulation and public concerns. British Medical Bulletin. 2000;**56**:62–73. DOI: 10.1258/0007142001902978

[118] Purchase IFH. What determines the acceptability of genetically modified food that can improve human nutrition? Toxicology and Applied Pharmacology. 2005;**207**:S19–S27. DOI: 10.1016/j.taap.2004.12.025

[119] Zuidmeer L, Goldhahn K, Rona RJ, Gislason D, Madsen C, Summers C, Sodergren E, Dahlstrom J, Lindner T, Sigurdardottir ST, McBride D, Keil T. The prevalence of plant food allergies: A systematic review. Journal of Allergy and Clinical Immunology. 2008;**121**:1210–1218. DOI: 10.1016/j.jaci.2008.02.019

[120] Naegeli H, Bresson J-L, Dalmay T, Dewhurst IC, Epstein MM, Firbank LG, Guerche P, Hejátko J, Moreno FJ, Mullins E, Nogué F, Sánchez Serrano JJ, Savoini G, Veromann E, Veronesi F, Casacuberta JM, Gennaro A, Paraskevopoulos K, Raffaello T, Rostoks N. Applicability of the EFSA Opinion on site-directed nucleases type 3 for the safety assessment of plants developed using site-directed nucleases type 1 and 2 and oligonucleotide-directed mutagenesis. EFSA Panel on Genetically Modified Organisms (EFSA GMO Panel). EFSA Journal. 2020;**18**:6299. DOI: 10.2903/J.EFSA.2020.6299

[121] Cabrera-Ponce JL, Barraza A, Alvarez-Venegas R. Cisgenic crops: Major strategies to create cisgenic plants based on genome editing. In Chaurasia A, Kole C, editors. Cisgenic

Crops: Potential and Prospects. Concepts and Strategies in Plant Sciences. Cham: Springer; 2022. p. 213–235. DOI: 10.1007/978-3-031-06628-3_11

[122] Ghose K, Yuan N, Dampanaboina L, Mendu V. Cisgenesis in the era of genome editing and modern plant biotechnology. In Chaurasia A, Kole C, editors. Cisgenic Crops: Potential and Prospects. Concepts and Strategies in Plant Sciences. Cham: Springer; 2022. p. 257–279. DOI: 10.1007/978-3-031-06628-3_13

[123] Gupta V, Sengupta M, Prakash J, Tripathy BC. Biosafety and bioethics. In Gupta V, Sengupta M, Prakash J, Tripathy BC, editors. Basic and Applied Aspects of Biotechnology. Media Singapore: Springer Science+Business; 2017. p. 503–520. DOI: 10.1007/978-981-10-0875-7_24

[124] Nouman Tahir M, Zahra S. Regulatory, biosafety, and ethical perspectives of plant genome editing. In Al-Khayri JM, Sattar MN, Sopory SK, Jain SM, editors. Genome Editing for Crop Improvement: Theory and Methodology. Wallingford, UK: CABI Books, CABI; 2025. p. 262–273. DOI: 10.1079/9781800622517.0018

[125] Shantharam S Regulatory review of transgenic plants: A perspective of the US experience. In: Ethics and equity in conservation and use of genetic resources for sustainable food security. Proceedings of a workshop to develop guidelines for the CGIAR, 21-25 April 1997, Foz do Iguacu, Brazil, IPGRI. Pp. 37–39. Available from: https://hdl.handle.net/10947/498

[126] Kuiper HA, Davies HV. The safe foods risk analysis framework suitable for GMOs? A case study. Food Control. 2010;**21**:1662–1676. DOI: 10.1016/j.foodcont.2010.02.011

[127] Hopp JZ. Contrasting directionalities in responsible innovation, lessons from the salmon farming industry in Chile and Norway. Technology in Society. 2025;**82**:102946. DOI: 10.1016/j.techsoc.2025.102946

[128] Hautea RA Global Status of Genetically Modified Crops: Current Trends and Prospects. In: Food for the Future: Opportunities for a Crowded Planet. Parliament House, Canberra, Australia: Crawford Fund for International Agricultural Research; 2002. p. 24–31. DOI: 10.22004/AG.ECON.123928

[129] Foley JK, Michaux KD, Mudyahoto B, Kyazike L, Cherian B, Kalejaiye O, Ifeoma O, Ilona P, Reinberg C, Mavindidze D, Boy E. Scaling up delivery of biofortified staple food crops globally: Paths to nourishing millions. Food and Nutrition Bulletin. 2021;**42**:116–132. DOI: 10.1177/0379572120982501

[130] Napier JA, Haslam RP, Tsalavouta M, Tsalavouta M, Sayanova O. The challenges of delivering genetically modified crops with nutritional enhancement traits. Nature Plants. 2019;**5**:563–567. DOI: 10.1038/S41477-019-0430-Z

[131] Janick J Horticultural plant breeding: Past accomplishments, future directions. In: International Symposium on Harnessing the Potential of Horticulture in the Asian-Pacific Region. Acta Horticulturae 2004; **294**: 61–65. DOI: 10.17660/ActaHortic.2005.694.6

[132] Brookes G, Barfoot P Crops: Global socio-economic and

environmental impacts 1996–2015. Dorchester: PG Economics Ltd; p. 1–201. 2017. Available from: https://www.bezpecnostpotravin.cz/UserFiles/publikace/2017globalimpactstudy.pdf [Accessed: 2025-October-05]

[133] Heap I. Global perspective of herbicide-resistant weeds. Pest Management Science. 2014;**70**: 1306–1315. DOI: 10.1002/ps.3696

[134] Varah A, Ahodo K, Coutts SR, Hicks HL, Comont D, Crook L, Hull R, Neve P, Childs DZ, Freckleton RP, Norris K. The costs of human-induced evolution in an agricultural system. Nature Sustainability. 2020;**3**:63–71. DOI: 10.1038/s41893-019-0450-8

[135] Wang S, Just DR, Pinstrup-Andersen P. Bt cotton and secondary pests. International Journal of Biotechnology. 2008;**10**:113–121. DOI: 10.1504/IJBT.2008.018348

[136] Pandian S, Ramesh M. Development of pesticide resistance in pests: A Key Challenge to the Crop Protection and Environmental Safety. In Srivastava PK, Singh VP, Singh A, Tripathi DK, Singh S, Prasad SM, Chauhan DK, editors. Pesticides in Crop Production. Wiley online Library; 2020. p. 1–13. DOI: 10.1002/9781119432241.ch1

[137] Van den Berg J, Hilbeck A, Bøhn T. Pest resistance to Cry1Ab Bt maize: Field resistance, contributing factors and lessons from South Africa. Crop Protection. 2013;**54**:154–160. DOI: 10.1016/j.cropro.2013.08.010

[138] Meissle M, Naranjo SE, Romeis J. Database of non-target invertebrates recorded in field experiments of genetically engineered Bt maize and corresponding non-Bt maize. BMC Research Notes. 2022;**15**:199. DOI: 10.1186/s13104-022-06021-3

[139] Jacobsen SE, Sørensen M, Pedersen SM, Weiner J. Feeding the world: Genetically modified crops versus agricultural biodiversity. Agronomy for Sustainable Development. 2013;**33**:651–662. DOI: 10.1007/s13593-013-0138-9

[140] Raney T, Pingali P. Private research and public goods: Implications of biotechnology for biodiversity. In Cooper J, Lipper LM, Zilberman D, editors. Agricultural Biodiversity and Biotechnology in Economic Development. Natural Resource Management and Policy. Vol. **27**. Boston, MA: Springer; 2005. p. 39–59. DOI:10.1007/0-387-25409-9_3

[141] Ryffel GU. Transgene flow: Facts, speculations and possible countermeasures. GM Crops and Food. 2014;**5**:249–258. DOI: 10.4161/21645698.2014.945883

[142] Garcia MA, Altieri MA. Transgenic crops: Implications for biodiversity and sustainable agriculture. Bulletin of Science, Technology and Society. 2005;**25**:335–353. DOI: 10.1177/0270467605277293

[143] Keese P. Risks from GMOs due to horizontal gene transfer. Environmental Biosafety Research. 2008;**7**:123–149. DOI: 10.1051/ebr:2008014

[144] Caradus JR. Is co-existence and/or containment of genetically modified plants possible, and is it important? New Zealand Journal of Agricultural Research. 2025;**68**:1329–1380. DOI: 10.1080/00288233.2025.2479724

[145] Paarlberg R. Starved for Science: How Biotechnology Is Being Kept Out of

Africa. Harvard: Harvard University Press; 2009. p. 93

[146] Maina J Africa's quiet biotech revolution: Gene editing emerges from Europe's anti-GMO shadow. Science for Sustainable Agriculture. 2025; Available from: https://www.scienceforsustainableagriculture.com/josephmaina2 [Accessed: 2025-October-06]

[147] Azadi H, Samiee A, Mahmoudi H, Jouzi Z, Rafiaani Khachak P, De MP, and Witlox F. Genetically modified crops and small-scale farmers: Main opportunities and challenges.. Critical Reviews in Biotechnology. 2015;**36**:434–446. DOI: 10.3109/07388551.2014.990413

[148] Louwaars N. Ethics watch: Controls over plant genetic resources — A double-edged sword. Nature Reviews Genetics. 2006;7:241. DOI: 10.1038/nrg1833

[149] Gerstetter C, Görlach B, Neumann K, Schaffrin D. The international treaty on plant genetic resources for food and agriculture within the current legal regime complex on plant genetic resources. The Journal of World Intellectual Property. 2007;**10**:259–283. DOI: 10.1111/j.1747-1796.2007.00323.x

[150] Caradus JR, Tumilson CA, Lin JY Intellectual property protection – Stimulating or constraining innovation and technology transfer? Joint X1 International Rangeland Congress and XXIV International Grassland congress (IRC/IGC) 2021. Contributed poster paper. P. 987–991. https://uknowledge.uky.edu/igc/24/7-2/10/

[151] Tokar B. The GMO threat to food sovereignty: Science, resistance and transformation. In Schanbacher E, editor. The Global Food System: Issues and Solutions. Connecticut, USA: Praeger; 2014. p. 173–190

[152] Shiva V. Stolen Harvest: The Hijacking of the Global Food Supply. Lexington, USA: University Press of Kentucky; 2016. p. 152. DOI: 10.2307/j.ctt19dzdd6

[153] Brookes G, Barfoot P. Farm income and production impacts of using GM crop technology 1996–2016. GM Crops and Food. 2018;**9**:59–89. DOI: 10.1080/21645698.2018.1464866

[154] Ortiz R. Critical role of plant biotechnology for the genetic improvement of food crops: Perspectives for the next millennium. Electronic Journal of Biotechnology. 1998;**1**:16–17. DOI: 10.4067/S0717-34581998000300005

[155] Herdt RW, Nelson R. Chapter 1 Biotechnology and Agriculture: Current and Emerging Applications. In Carter CA, Moschini GC, Sheldon I editors. Genetically Modified Food and Global Welfare. UK: Emerald Group Publishing Limited; 2011. DOI: 10.1108/S1574-8715

[156] WUR. WUR gives away CRISPR intellectual property licenses for free in fight against hunger. 2022; Available from: https://www.wur.nl/en/newsarticle/WUR-gives-away-CRISPR-intellectual-property-licenses-for-free-in-fight-against-hunger.htm#:~:text=Today%20Wageningen%20University%20%26%20Research%20%28WUR%29%20announces%20it,applied%20to%20gene-editing%20of%20plants%20for%20non-profit%20applications [Accessed: 2025-October-04]

[157] Li J, Jiggins J, ETL LVB, Leeuwis C. Towards a regime change in

the organization of the seed supply system in China. Experimental Agriculture. 2013;**49**:114–133. DOI: 10.1017/S001447971200097X

[158] Li J, Leeuwis C, ETL LVB, Song Y, Jiggins J. Contribution of action researching to institutional innovation: A case study of access and benefit sharing (ABS) mechanisms in the participatory plant breeding (PPB) in Southwest China. International Journal of Agricultural Resources, Governance and Ecology. 2012;**9**:204–227. DOI: 10.1504/IJARGE.2012.050349

[159] Clapp J. The problem with growing corporate concentration and power in the global food system. Nature Food. 2021;**2**:404–408. DOI: 10.1038/s43016-021-00297-7

[160] Clapp J. The rise of big food and agriculture: Corporate influence in the food system. In Sage C, editor. A Research Agenda for Food Systems. Cheltenham, UK: Edward Elgar Publishing; 2022. p. 45–66. DOI: 10.4337/9781800880269.00011

[161] Bonny S. Corporate concentration and technological change in the global seed industry. Sustainability. 2017;**9**: Article1632. DOI: 10.3390/su9091632

[162] De Schutter O. The political economy of food systems reform. European Review of Agricultural Economics. 2017;**44**:705–731. DOI: 10.1093/erae/jbx009

[163] Daviron B, Douillet M Major players of the international food trade and the world food security. Research Papers in Economics. FOODSECURE Working papers 12, LEI Wageningen UR. 2013. Available from: https://ideas.repec.org/p/fsc/fspubl/12.html [Accessed: 2025-November-22]

[164] Naylor RL. A global perspective on food systems. Journal of Agriculture, Food Systems, and Community Development. 2015;**5**:15–18. DOI: 10.5304/jafscd.2015.052.013

[165] Raimi L, Masri M. Critical discourse on GMO agripreneurship for global food security: Bridging agricultural colonialism and food sovereignty perspectives. International Journal of Ethics and Systems. 2025. DOI: 10.1108/IJOES-02-2025-0079

[166] Sundar MS, Vezhaventhan D, Archana ARA. Tracing U.S. influence in the construction of the global food regime. Discover Food. 2025;**5**:316. DOI: 10.1007/s44187-025-00627-9

[167] Smyth SJ, Falck-Zepeda J, Ludlow K. The costs of regulatory delays for genetically modified crops. Journal of International Law and Trade Policy. 2016;**17**:173–195. DOI: 10.22004/AG.ECON.253074

[168] Morris M, Edmeades G, Pehu E. The global need for plant breeding capacity: What roles for the public and private sectors? HortScience. 2006;**41**:30–39. DOI: 10.21273/HORTSCI.41.1.30

[169] Hubbell BJ, Welsh R. Transgenic crops: Engineering a more sustainable agriculture? Agriculture and Human Values. 1998;**15**:43–56. DOI: 10.1023/A:1007437922114

[170] Rifkin J. The Biotech Century: The Coming Age of Genetic Commerce. London: Penguin Publishing Group; 1999. p. 272

[171] Mueller NG, Flachs A. Domestication, crop breeding, and genetic modification are fundamentally different processes: Implications for

seed sovereignty and agrobiodiversity. Agriculture and Human Values. 2022;**39**:455–472. DOI: 10.1007/s10460-021-10265-3

[172] Duvick DN Interactions of CGIAR and private industry: An overview of ethical issues and equity concerns. In: Ethics and equity in conservation and use of genetic resources for sustainable food security. Proceedings of a workshop to develop guidelines for the CGIAR, 21-25 April 1997, Foz do Iguacu, Brazil, IPGRI. Pp. 139–145. https://hdl.handle.net/10947/498

[173] Louwaars NP, Tripp R, Eaton D, Henson-Apollonio V, Hu R, Mendoza M, Muhhuku F, Pal S, Wekundah J Impacts of strengthened intellectual property rights regimes on the plant breeding industry in developing countries. Report commissioned by the World Bank. Wageningen, The Netherlands. 2005. Pp. 176. Available from: https://www.iprsonline.org/resources/docs/Louwaars CGN_Plants_05.pdf [Accessed: 2025-October-04]

[174] Coffman WR, Lesser WH, McCouch SR Commercialization and the scientific research process: The example of plant breeding. Science and the University. 2003. Available from: https://archive.ilr.cornell.edu/sites/default/files/Commercialization%20and%20the%20Scientific%20Research%20Process.pdf [Accessed: 2025-October-04]

[175] Beard R End-Point versus Point of Sale levying of plant breeding royalties: An economic analysis using optimal control theory. Munich Personal RePEc Archive (MPRA) Paper No. 7178. Pp. 22. 2008. Available from: https://mpra.ub.uni-muenchen.de/7178/ [Accessed: 2025-October-04]

[176] Kingwell R, Watson A. End-point royalties for plant breeding in Australia. Agenda. 1998;**3**:323–334. Available from: https://www.jstor.org/stable/43198977

[177] Brummer EC, Barber WT, Collier SM, Cox TS, Johnson R, Murray SC, Olsen RT, Pratt RC, Thro AM. Plant breeding for harmony between agriculture and the environment. Frontiers in Ecology and the Environment. 2011;**9**:561–568. DOI: 10.1890/100225

[178] Edgerton MD. Increasing crop productivity to meet global needs for feed, food, and fuel. Plant Physiology. 2009;**149**:7–13. DOI: 10.1104/pp.108.130195

[179] Fussell GE. Farming Techniques from Prehistoric to Modern Times. New York: Pergamon Press; 1966. p. 181

[180] Hossain F, Onyango BM, Schilling BJ, Hallman WK. Public perceptions of biotechnology and acceptance of genetically modified food. Journal of Food Distribution Research. 2003;**34**:36–50. DOI: 10.22004/ag.econ.27047

[181] Cabelkova I. Understanding public perception of genetically modified food: Navigating misinformation and trust. Food Science and Nutrition Technology. 2024;**9**:354. DOI: 10.23880/fsnt-16000354

[182] Landrum AR, Hilgard J, Lull RB, Akin H, Jamieson KH. Open and transparent research practices and public perceptions of the trustworthiness of agricultural biotechnology organizations. Journal of Science Communication. 2018;**17**:A04. DOI: 10.22323/2.17020204

[183] Straughan R. Ethics, morality and crop biotechnology. 2. Extrinsic concerns. Outlook on Agriculture. 1995b;**24**:233–240. DOI: 10.1177/003072709502400407

[184] Carr S, Levidow L. How biotechnology separates ethics from risk. Outlook on Agriculture. 1997;**26**:145–150. DOI: 10.1177/003072709702600303

[185] Levin M. The role of risk assessment in developing statutes and regulations. In Krattiger AF, Rosemarin A, editors. Biosafety for Sustainable Agriculture. ISAAA: Ithaca and SEI. Stockholm: ISAAA; 1994. p. 127–138. https://www.isaaa.org/resources/publications/downloads/biosafety_bk.pdf#page=145

[186] Robinson J. Ethics and transgenic crops: A review. Electronic Journal of Biotechnology. 1999;**2**(2). Available from: https://utoronto.scholaris.ca/server/api/core/bitstreams/246b2874-157c-4b1c-b96b-fe86d93c0e58/content

[187] Levidow L, Carr S. How biotechnology regulation sets a risk/ethics boundary. Agriculture and Human Values. 1997;**14**:29–43. DOI: 10.1023/A:1007394812312

[188] Norwegian Directorate for Nature Development. The Norwegian Gene Technology Act and socio-economic. 2011. Available from: https://www.miljodirektoratet.no/globalassets/dokumenter/publikasjoner/faktaark/faktaark-gmo-sosiookonomi.pdf [Accessed: 2025-October-02]

[189] Rolston H. Justifying sustainable development: A continuing ethical search. Global Dialogue (Centre for World Dialogue, Nicosia, Cyprus). 2002;**4**(1):103–113. Available from: https://api.mountainscholar.org/server/api/core/bitstreams/7abe4e6a-c06d-4c9b-8ec7-b0c63ab9b5ff/content

[190] Simona M The ethical dimension of the sustainable development. Proceedings of the 8th International Management Conference "Management Challenges for Sustainable Development", 2014, Bucharest, Romania. Pp. 818–825. https://conferinta.management.ase.ro/archives/2014/pdf/81.pdf

[191] Verma A. Sustainable development and environmental ethics. International Journal on Environmental Sciences. 2019;**10**:1–5. https://papers.ssrn.com/sol3/papers.cfm?abstract_id=3689046

[192] Callahan D Setting and communicating the limits in biotechnology. In: Biotechnology and Ethics: A Blueprint for the Future. 1996. Available from: https://www.iatp.org/sites/default/files/Setting_and_Communicating_the_Limits_in_Biotec.htm [Accessed: 2025-October-06]

[193] Verma SR. Genetically modified plants: Public and scientific perceptions. International Scholarly Research Notices Biotechnology. 2013;**2013**(1):820671. DOI: 10.5402/2013/820671

[194] Westengen OT, Dalle SP, Mulesa TH Navigating toward resilient and inclusive seed systems. Proceedings of the National Academy of Sciences of the United States of America. 2023; **120**: e2218777120. DOI: 10.1073/pnas.2218777120

[195] Marchant GE, Stevens YA. A new window of opportunity to reject

process-based biotechnology regulation. GM Crops and Food. 2015;**6**:233–242. DOI: 10.1080/21645698.2015.1134406

[196] Qaim M. Genetically Modified Crops and Agricultural Development. Basingstoke, UK: Palgrave Macmillan; 2016. p. 206p.ISBN:9781137405722

[197] Yadav V, Pal D, Poonia AK. A study on genetically engineered foods: Need, benefits, risk, and current knowledge. Cell Biochemistry and Biophysics. 2024;**82**:1931–1946. DOI: 10.1007/s12013-024-01390-x

[198] Carracelas G, Guilpart N, Cassman KG, Grassini P. Distinguishing between yield plateaus and yield ceilings: A case study of rice in Uruguay. Field Crops Research. 2023;**292**:108808. DOI: 10.1016/j.fcr.2023.108808

[199] Chapman DF, Mackay AD, Caradus JR, Clark DA. Pasture productivity in New Zealand: Trends, factors, and prospects. New Zealand Journal of Agricultural Research. 2024. DOI: 10.1080/00288233.2024.2425071

[200] Herrera JM, Levy Häner L, Mascher F, Hiltbrunner J, Fossati D, Brabant C, Charles R, Pellet D. Lessons from 20 years of studies of wheat genotypes in multiple environments and under contrasting production systems. Frontiers in Plant Science. 2020;**10**:1745. DOI: 10.3389/fpls.2019.01745

[201] Moore FC, Lobell DB The fingerprint of climate trends on European crop yields. Proceedings of the National Academy of Sciences of the United States of America. 2015; **112**: 2670–2675. DOI: 10.1073/PNAS.1409606112

[202] Thorrold BS. Where to now for forage productivity in dairying? Journal of New Zealand Grasslands. 2024;**86**: 33–35. DOI: 10.33584/jnzg.2024.86.3712

[203] Zampieri M, Toreti A, Ceglar A, Naumann G. Climate resilience of the top ten wheat producers in the Mediterranean and the Middle East. Regional Environmental Change. 2020;**20**:41. DOI: 10.1007/s10113-020-01622-9

[204] Institute of Medicine. National Academy of Sciences, and National Academy of Engineering. On Being a Scientist: Responsible Conduct in Research. Second. Washington, DC: The National Academies Press. p. 40; 1995. DOI: 10.17226/4917

[205] Wagner FH. Ethics, science, and public policy. BioScience. 1996;**46**: 765–766. DOI: 10.2307/1312852

Chapter 5

IPRs, Bioethics, and Equity: Advancing Genome Editing in Africa's Bio-Economy

Muyiwa S. Adegbaju, Olalekan Akinbo, Rachel Shibalira, Olufunke B. Shittu, Ruth Asheazi Manzah, Sunday Igu Rock Akile, Charles Oluwaseun Adetunji and Bolaji N. Thomas

Abstract

Genome editing (GEd) technologies present a potent avenue for addressing critical challenges in Africa, spanning agriculture and human health. However, their full potential is hindered by significant financial and infrastructural difficulties, which obscure the underlying systemic obstacles to successful implementation. This paper argues that the success of GEd in Africa is constrained by a dual challenge rooted in the continent's fragmented governance and capacity systems: the innovation-to-market gap and the monopolistic risk. The innovation-to-market gap stems from a market failure caused by ineffective intellectual property rights (IPR), which discourage private investment, compounded by a lack of commercial focus in public research institutions. Simultaneously, the monopolistic risk poses a threat of foreign control over vital technologies due to reliance on external entities for licensing key patents. Overcoming these obstacles necessitates moving beyond a singular focus on IPR and instead adopting a truly holistic bio-economy policy. This policy must be founded on four interconnected pillars: strengthening IPR frameworks, building local technical and infrastructural capacity, ensuring robust bioethical governance, and promoting equitable access to GEd technologies like CRISPR-Cas9. We conclude that only by addressing these systemic failures – spanning legal, economic, and institutional domains – can GEd truly thrive and support Africa's long-term development goals.

Keywords: intellectual property rights frameworks, genome editing, bio-economy policy, Africa, CRISPR-Cas9, monopolistic risk, innovation-to-market gap

1. Introduction

African nations are strategically prioritizing the development of a resilient, self-sufficient, and sustainable agricultural and resource management system.

Rapid population growth drives this need, placing unprecedented pressure on existing ecosystems and demanding strategies that support sustainable resource management and long-term ecological health [1]. This commitment has led to the creation of a transformative continental blueprint known as the bioeconomy [2]. This framework provides the foundation for meeting local needs for food, energy, and sustainable goods, making it the primary focus for addressing ongoing challenges like malnutrition and food insecurity [3, 4]. Advances in biotechnology are the key drivers of the bioeconomy. These advances include developing life-saving interventions for diseases like COVID-19 and malaria, as well as creating resilient, nutritious, and high-yield crop varieties [5–11]. Among these tools, genome editing (GEd) stands out as a powerful opportunity to transform African food systems, improve public health, and secure a prosperous future [12].

Achieving this potential requires a well-designed, goal-oriented policy framework that governs innovation and technology transfer. Its importance is underscored by the fragmented state of existing innovation capacity across the continent. To bridge the gap between GEd's immense potential and Africa's current situation, it's essential to first identify the main weaknesses in scientific and technological innovation capacity. The 2025 Global Innovation Index (GII), compiled by the World Intellectual Property Organization (WIPO), provides an objective way to measure a country's innovation performance (**Figure 1**). Most African countries score low on this index – Nigeria (105th) and South Africa (61st) among them – compared to global leaders such as Switzerland (1st) and Sweden (2nd). This gap clearly underscores the institutional and market shortcomings that define the continent's dual IPR challenge. This challenge involves two related risks: the first is weak domestic commercialization, characterized by an underdeveloped innovation environment and limited local capacity to turn innovations into market-ready products. A lack of strong IPR protections hinders the commercialization of local inventions, depriving African scientists and entrepreneurs of incentives to invest in costly processes from research to market entry. This stifles the growth of a vibrant bioeconomy and keeps promising research confined to academic circles.

The second risk concerns ethical issues regarding foreign IPR monopolies. Low GII rankings make Africa vulnerable to external IPR regimes that could monopolize key transformative technologies. For instance, the ongoing legal battle over the ownership of foundational tools such as CRISPR-Cas9 [13] illustrates how monopolistic control can restrict access to crucial research and development. Such restrictions would drive up costs and limit the availability of essential goods – like disease-resistant crops and vaccines – posing risks to food security and public health. The combined legal, institutional, and financial shortcomings form the core of Africa's dual bio-economic challenge: policies must attract private investment through strong IPR protections while ensuring public access to vital technologies. To navigate this complex landscape, this chapter adopts the Policy-to-Action Framework as its primary analytical tool. We contend that a balanced policy approach is critical: it must create a conducive environment for local innovators with strong incentives while allowing flexibility to ensure public access to key biotechnologies. By analyzing this dual challenge and proposing a four-pillar policy-to-action framework, this review offers a sustainable path toward Africa's scientific sovereignty and economic growth.

DOI: http://dx.doi.org/10.5772/intechopen.1014542

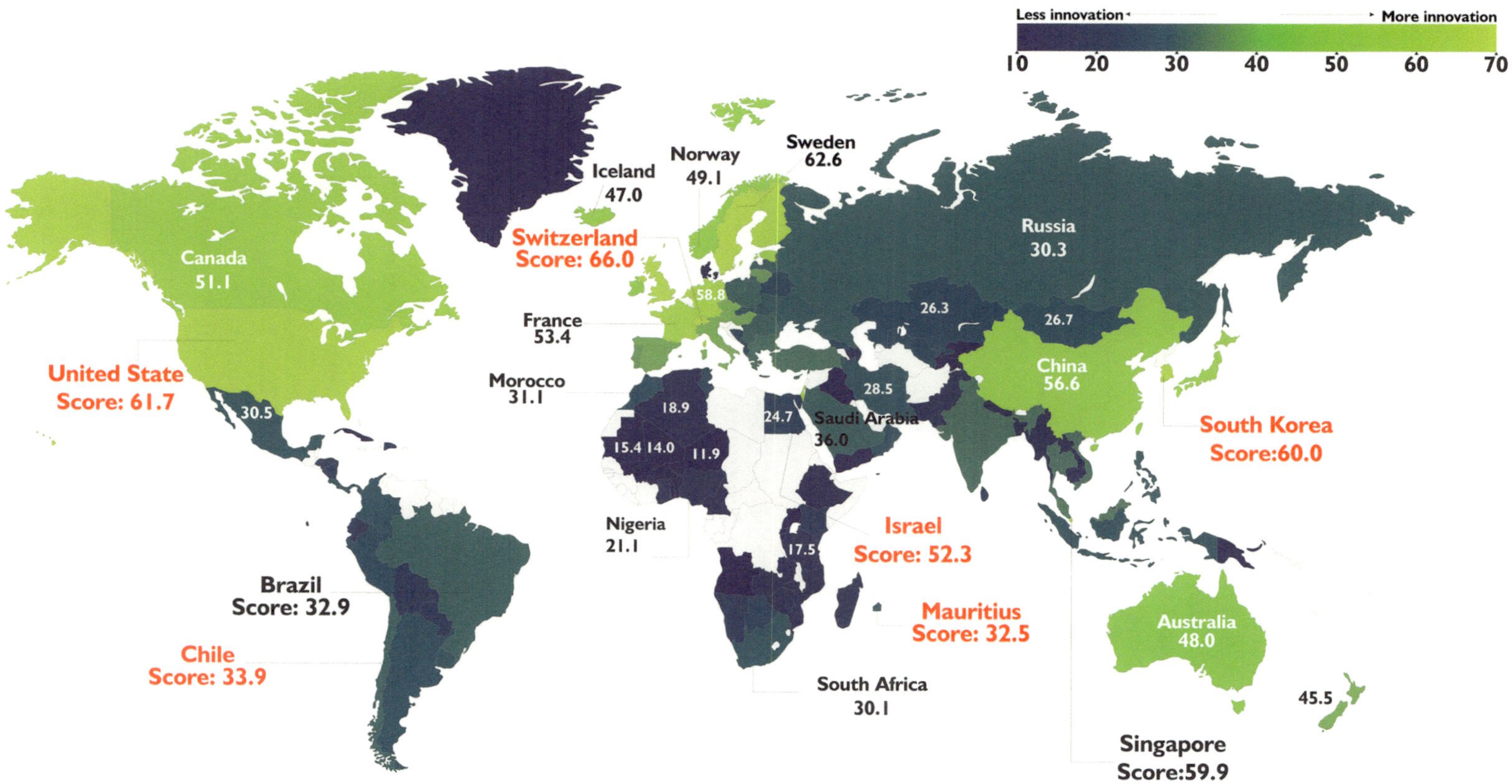

Figure 1.
Global Innovation Index (GII) ranking, 2025 [14]. The map displays the innovation performance of 139 countries, as measured by the GII compiled by the World Intellectual Property Organization (WIPO). Rankings are based on seven key pillars: Institutions, human capital & research, infrastructure, market sophistication, business sophistication, knowledge & technology outputs, and creative outputs. Countries highlighted in red represent the highest-ranking innovators within their respective regions: North America (United States, 3rd), Europe (Switzerland, 1st), the Middle East (Israel, 17th), Asia (Republic of Korea, 4th), Africa (Mauritius, 53rd), and South America (Chile, 42nd). Countries with no available data are shown in gray. Notable African improvements include Morocco (57th, up nine ranks) and Namibia (91st, up 11 ranks), with Sub-Saharan Africa featuring six overperformers.

2. The dual challenge of IPR in African genome editing

Genome editing holds great promise for advancing Africa's bioeconomy, a view widely supported by experts and policymakers [15, 16]. The optimism largely comes from GEd's pioneering agricultural applications, which have led to calls for new regulatory frameworks. Such regulatory frameworks often aim to distinguish GEd products from genetically modified organisms (GMOs) and to adopt less stringent oversight to expedite adoption across the continent [17]. However, the scientific potential is immediately complicated by infrastructural realities. Advances in genome sequencing have further boosted GEd's theoretical potential, with costs dropping globally from roughly USD 100 million in 2001 to just over USD 500 in 2023 [18]. Despite this global trend, significant regional disparities remain. For instance, in Africa, sequencing costs are 2–3 times higher than in the United States due to import tariffs, reagent shortages, and logistical issues, with sample shipping expenses reaching up to USD 4,500.

The quoted figures represent only material costs, excluding equipment, labor, and maintenance. Such persistent infrastructural and financial difficulties obscure the significant technical, bioethical, and institutional challenges involved in successful GEd implementation [19]. By implication, realizing GEd's full potential demands more than capital; it requires well-funded, equipped labs, expertise in complex genetic targeting, and the ability to manage third-party licensing and international regulations [20]. These essential demands highlight that the success of GEd in Africa depends on overcoming two connected obstacles rooted in the continent's fragmented governance and capacity systems: the innovation-to-market gap – transforming scientific discoveries into tangible commercial products – and the monopolistic risk – the threat of foreign control over vital technologies. Addressing the challenge posed by these risks is crucial for GEd to truly thrive and support Africa's development. Therefore, ensuring that GEd serves Africa's development goals requires a truly holistic bio-economy policy that addresses IPR, bioethics, and equity by overcoming the fundamental infrastructural barriers hindering fair access and sustainable commercialization.

2.1 The innovation-commercialization gap: Fragmentation and market failure

Fulfilling this holistic mandate first necessitates a deep analysis of the two connected obstacles that limit Africa's self-sufficiency: the innovation-to-market gap and the monopolistic risk. A strong and transparent IPR framework is crucial for turning scientific breakthroughs into market-ready products. While many GEd projects are starting across Africa [12], the high-risk biotech sector needs legal security for investment, which is often seen as lacking. In this field, investment depends heavily on well-defined and protected IPR. This legal security is vital because it grants exclusive rights, boosts a company's value, and reduces the risk of competitors copying innovations. Without effective IPR to secure an exclusive return, private capital – both domestic and international – lacks the fundamental incentive to fund the large, long-term development of GEd products, thereby leading directly to market failure. This failure stems from deep structural flaws in the legal and technical systems governing IP, creating a continuous cycle of underinvestment.

The ongoing cycle of underinvestment often prevents local genome-editing research from moving beyond the laboratory. Despite broad institutional challenges reflected by low GII rankings (**Figure 1**), country-specific studies from Ghana, Namibia, and Uganda show that scientific innovation exists, but the commercial pathway is largely broken. For example, national research institutions have developed complex bio-based solutions – such as Uganda's National Agricultural Research Organisation's (NARO) work on an anti-tick vaccine – but the critical commercial support needed to scale the bioeconomy is lacking. This is compounded by the fact that securing patents and developing revenue-generating products are often not top priorities for these institutions, as shown by a survey of health research institutions across 42 sub-Saharan African countries [21]. Consequently, the lack of commercial focus exacerbates the chronic underinvestment problem, as the private sector perceives the absence of a patent-driven institutional culture as a major financial risk. Creating a serious disconnect between the public R&D sector's mission and the creation of commercial value, this problem is compounded by the fact that Africa's weak link to the global market is twofold: a failure to commercialize research from local public institutions and a reliance on external entities for licensing key technologies.

As shown in **Table 1**, low IP indices correlate with lower FDI and R&D, thereby exacerbating these gaps.

The history of Ghana's Ayensu Starch Factory, which processes cassava (a primary GEd target), clearly shows the real-world impact of this commercialization bottleneck. The company's early years were marked by high operational costs and two temporary shutdowns, highlighting the initial fragility of bio-based value chains. Importantly, the subsequent recovery was achieved only through sustained, targeted public policy efforts – specifically, the government's introduction of an excise duty waiver, leading to exclusive supply agreements. This history confirms that building commercial resilience for bioeconomy outputs requires not only innovation but also ongoing, strategic public policies like strong IPR to close the commercialization gap. Such a structural gap has prompted targeted government actions, such as the planned Uganda Biosciences Park initiative, designed to address this missing infrastructure.

Country/ Region	IP index score (2025)	Key economic impact statistic	Source correlation with growth
Nigeria	34.28%	80% software piracy; unicorn valuation leads globally.	1% IP strength → 2.8% FDI rise (OECD)
South Africa	35.15%	32% piracy; patents positively explain GDP growth.	Digitalization + patents boost growth (study).
Kenya	36.68%	Online piracy costs $750 M (2022).	Weak enforcement hinders R&D (0.7% potential rise).
Sub-Saharan Africa	~41.82% (regional avg.)	450% trademark growth post-TRIPS	Strong IP → increased trade, jobs in IP-intensive sectors
Morocco	59.21%	High-tech manufacturing 50% of output	Overperformer; IP incentives drive value-added production

Table 1.
Qualitative impacts of IPR strength on economic indicators in the african context [22].

2.2 Policy dilemmas and legal uncertainty

Africa has demonstrated a clear intention to strengthen its legal infrastructure, as shown by the ratification of the convention establishing the WIPO by all 54 countries [23], and by essential harmonization efforts by regional organizations like the African Regional Intellectual Property Organization (ARIPO) and the African Intellectual Property Organization (OAPI) [24]. The main challenge, however, remains the inconsistent application and weak enforcement of these emerging frameworks. This systemic legal instability actively discourages the necessary private investment from entering the high-risk biotechnology sector. Addressing this instability requires policymakers to balance compliance with international standards, such as the TRIPS Agreement, with crucial national goals, including food security. The inability to effectively manage this core policy dilemma perpetuates legal instability, restricts capital flows, and complicates the management of the IPR nexus – the intersection of IP law, sectoral regulation, and national development policy – within national systems. The IPR nexus involves two challenging institutional and ethical dimensions that must be reconciled. First, with respect to institutional complexity, regulating the bioeconomy requires extensive cross-sector coordination among government agencies, as evidenced by policies such as Uganda's Biofuels Act; therefore, a cohesive IPR framework must be integrated into this fragmented institutional landscape. Second, with respect to ethical considerations and the protection of knowledge, policies such as Uganda's National IP Policy explicitly emphasize safeguarding traditional knowledge (TK) [25]. Since genome editing often targets traits found in native species, this focus is vital to ensure strong protections are in place to prevent external entities from gaining monopolistic control over vital indigenous resources.

2.3 Strategic trade-offs in IPR implementation

Attempting to navigate this IPR nexus – balancing global IP standards with local development needs – is inherently filled with strategic trade-offs. Strategies aimed at protecting national interests, such as using IPR flexibilities, can create legal uncertainties that discourage investment. Implementing IPR flexibilities – like compulsory licensing, which allows countries to adapt rules to their specific needs – poses a challenging balancing act. While using competition law to prevent monopolies in agricultural technologies is a strong public-interest tool, it also introduces significant legal uncertainty for foreign investors and requires specialized legal and institutional expertise for effective enforcement against multinational corporations. Moreover, establishing unique (sui generis) systems for plant variety protection, intended to safeguard farmers' rights, can lead to a less predictable IPR environment, thereby deterring private-sector breeders from investing in new genome-edited crop varieties. A persistent shortage of skilled personnel in IPR management, valuation, and enforcement worsens these strategic trade-offs. Without the legal and technical expertise to effectively handle these complex, high-stakes IP frameworks, the system remains unreliable for securing the substantial, long-term investments needed for successful genomic innovation. The combined effect of this structural uncertainty and the skills gap underscores the need for a policy approach that proactively guards against monopolistic risks while encouraging socially responsible licensing of key technologies, such as those dominating the CRISPR patent landscape [13].

2.4 Regional variations in IPR enforcement and the impact on GEd commercialization

The resulting systemic legal fragmentation, driven by policy trade-offs and the skills gap, significantly hinders the development of a strong GEd commercialization ecosystem. The inconsistent enforcement and rising risks across different regions are best understood through regional analysis that highlights varying levels of IPR maturity and distinct commercial challenges. Regions with more developed economies and legal frameworks, such as North Africa (e.g., Egypt and Morocco) and Southern Africa (e.g., South Africa), generally have IPR systems that align better with international standards. However, in South Africa, these systems are often fragmented across multiple laws and regulations, such as those governing patents, designs, and plant breeders' rights, which complicates their practical application [26, 27]. Even in these areas, the GEd commercialization pipeline faces two deep-rooted issues that go beyond basic IPR quality. First, regulatory misalignment persists because some GEd products are still classified as GMOs under current laws, leading to costly and lengthy regulatory processes [28]. Second, IPR protection becomes less effective outside urban centers, where enforcement is weak and public awareness is low, especially in rural and farming communities. This vulnerability makes patents for gene-edited seeds or livestock traits more susceptible to infringement at the local level, further undermining investor confidence in large-scale agricultural biotech projects. For instance, studies in South Africa have shown that basic knowledge of national IP laws is not being shared within farming communities and schools, highlighting a significant literacy gap that impedes effective local compliance and protection [29].

In economic powerhouses like West Africa (e.g., Nigeria and Ghana) and key East African states (e.g., Kenya and Tanzania), the main challenge isn't the laws themselves but their consistent implementation and strict enforcement. Although countries such as Ghana and Kenya align their IP laws with international standards, their judicial and technical capacity to handle the complexities of biotech patent litigation varies [30, 31]. This systemic gap is significant because studies in Ghana show that, despite having a basic IP system, issues such as weak enforcement, limited access, and low public awareness actively hinder innovative risk-taking and decrease investor confidence in the emerging startup economy [32]. This failure in implementation is particularly evident in Nigeria, where an emerging GEd research sector faces major obstacles from widespread piracy and weak judicial enforcement [20, 33], which greatly impede the monetization of local discoveries.

Critically, this region often shows a large gap between regulatory progress and commercial protection. Many countries, including Nigeria and Kenya, have made notable advances in adopting science-based biosafety regulations for genome-edited products [28]. This regulatory clarity shows government commitment and often reduces technical barriers to market entry. However, this progress in biosafety doesn't automatically lead to economic benefits because of underlying IPR issues. For emerging technologies like GEd, legal uncertainty caused by weak and fragmented IP enforcement undermines the positive effects of biosafety approvals. This legal uncertainty acts as a non-tariff barrier to investment [32], creating an unacceptable risk profile that continues to discourage vital venture capital, regardless of scientific readiness. This situation highlights the need for a comprehensive approach, where strong IPR systems and supportive biosafety frameworks work together to unlock the full commercial potential of GEd products.

2.5 Judicial precedents: Affirming IPR enforcement capacity in agricultural biotechnology

The previous analysis revealed that Africa's IPR landscape faces inconsistent enforcement and weak implementation, deterring investment. However, despite these systemic challenges, some judicial precedents from key economies show that the necessary enforcement tools and legal frameworks exist – they are just applied inconsistently. These case studies highlight a crucial point: they confirm the legal mechanisms that must be reliably used to boost investor confidence in the high-value biotechnology sector, providing essential proof of concept for protecting valuable genomic technologies. The first significant precedent for patent protection is seen in the outcome of Monsanto v. Cargill [34]. This landmark case, governed by South African patent law, involved Cargill's unauthorized use of Monsanto's patented Roundup Ready technology in soybeans. The South African High Court's 2017 decision in favor of Monsanto confirmed that Cargill had infringed the patent, ordering the company to pay damages and stop unauthorized use. This ruling has important implications for the genomics industry. It set a clear legal precedent showing that South Africa's courts are willing and able to enforce patents on advanced agricultural biotechnology products, even against large corporations. Such judicial consistency signals to both foreign and local investors that their exclusive rights will be protected – directly addressing one of the main barriers to capital flow identified in this analysis.

Beyond traditional patents, the legal protection of new crops also relies on sui generis systems like plant variety rights (PVR). In the 2002 case of Keith Kirsten's (Pty) Limited v Weltevrede Nursery (Pty) [35] Limited and Others, the Western Cape High Court ruled in favor of Keith Kirsten's, finding that Weltevrede Nursery had infringed on the plaintiff's PBR for the *Canna Phasion* variety – a pink-veined, burgundy-striped ornamental canna developed through mutation breeding. The court confirmed infringement under Section 2.3 of the Plant Breeders' Rights Act 15 of 1976, rejected defenses based on the variety's novelty and distinctiveness, and awarded statutory damages of R10,000 along with costs (Keith Kirsten's (Pty) Limited v Weltevrede Nursery (Pty) Limited and Others, 2002). This case highlights the enforceability of PVR, which are essential for protecting both traditional breeding outputs and modern techniques like genome editing. By effectively enforcing its PVR, Keith Kirsten's demonstrated that the legal system provides recourse for innovators investing in developing locally adapted, high-yielding traits. This judicial support for PVR is vital to incentivize ongoing plant breeding and secure investments in advanced innovations like genome editing (GEd), which are critical for Africa's food security.

2.6 Synthesis: IPR fragmentation as a non-tariff barrier

This detailed comparative analysis reveals a complex continental IPR landscape. While institutional support for robust IPR is evidently growing, the limited judicial precedents affirming the enforceability of high-value biotechnology and Plant Variety Rights are confined to only one jurisdiction. Specifically, successful patent and PVR cases showcase South Africa's legal capacity, yet this strength remains isolated and fragmented across the continent. Overall, the region continues to struggle with weak routine enforcement, widespread piracy, and a persistent lack of legal awareness. This systemic disunity poses a fundamental threat to the goal of continental economic integration. The widespread legal disunity not only establishes a de facto Non-Tariff Barrier (NTB) that actively obstructs the primary objectives of the African Continental Free Trade Area (AfCFTA), but also leads to a critical failure to secure the private capital necessary to bridge the gap from innovation to commercialization and to develop a regionally connected bio-economy. Therefore, two urgent actions are required: achieving regional harmonization and implementing intensive capacity building. Both strategies must be carefully designed to address the specific legal and technical needs of the genomic innovation sector. However, pursuing harmonization introduces an immediate and significant ethical concern: the policy tools employed to attract private investment may unintentionally grant foreign or domestic entities monopolistic control over vital life-saving and food-producing technologies.

2.7 The ethical dilemma of IPR monopolies and the privatization of research

The conflict between promoting private investment and ensuring public access is rooted in a fundamental change in the global funding landscape for agricultural and health research and development. This change demands legal mechanisms to create checks and balances against market dominance. Because of the heavy reliance on external sources of innovation, the ethical challenge faced by African policymakers concerning IPR comes directly from the widespread global dependence on the private sector to lead innovation. Global data confirms that the biotechnology sector is mainly financed by private interests, with only about 20% of investments coming from the public sector [36, 37]. This private dominance has increased since the mid-1990s as public funding for agriculture sharply declined [38]. This economic reality links directly to IPR requirements. Since the private sector is primarily driven by profit, its willingness to fund high-risk, long-term development of genome-edited products depends entirely on the legal assurance that its IP will be protected from copying [39]. Local studies in Ghana support this, showing that IP documents like patents and trademarks are key indicators of credibility and risk reduction, with investors often demanding IPR proof during negotiations and sometimes withholding funds until rights are secured. Furthermore, legal IP ownership provides critical bargaining power and credibility for local innovators involved in cross-border alliances, franchising, and licensing agreements [32]. A strong and enforceable IPR system is, therefore, not merely a policy goal but an essential prerequisite for attracting and retaining private capital and expertise in African biotechnology.

Consequently, as African nations are compelled to strengthen their IPR laws to attract necessary investment, they risk inadvertently granting foreign or domestic companies monopolistic control over crucial genome-editing technologies. This looming ethical challenge is driven by economic necessity, and policymakers must balance this private incentive with vital national development goals, especially food security. The global patent disputes over foundational solutions like CRISPR-

Cas9 and its variants clearly show how such monopolies can form, restricting fair access to innovation in developing markets and harming public welfare for private gain. This risk is heightened by the layered control of genomic technology. Monopolistic control often extends beyond one final product to include "upstream patents" on foundational tools, such as the Cas9 enzyme itself. Control over these upstream technologies allows patent holders to set the terms for all "downstream" innovations, which could include a locally developed drought-resistant cassava. Furthermore, overlapping patent rights can quickly create a complex patent thicket [40], leading to royalty stacking. This results in prohibitively high cumulative costs for local African innovators [41] just to determine Freedom-to-Operate (FTO), effectively blocking them from commercializing their work. Ultimately, a patented drought-resistant seed could force subsistence farmers to pay high, recurring fees, directly threatening national food security and developmental goals. The global patent disputes over foundational solutions such as CRISPR-Cas9 and its variants clearly highlight the potential for such monopolies to form, thereby limiting fair access to innovation in developing markets and jeopardizing public welfare for private gain.

To address diversity oversights in IPR frameworks, it is crucial to recognize gender intersections, where women remain underrepresented in patenting – accounting for only 15% of patent applications in Africa and 9% in South Africa – due to patriarchal biases that undervalue communal innovations like traditional crafts [18, 42]. This gap hampers women's economic empowerment and highlights the need for inclusive policies, such as gender-sensitive IP training and sui generis protections for traditional cultural expressions. Beyond traditional knowledge (TK) protections, indigenous knowledge in genome editing raises ethical concerns, such as colonial legacies of dispossession, power inequalities in partnerships, and the importance of community co-design to ensure fair benefits and prevent accidental misuse [43, 44]. For instance, proposals for gene-edited organisms should incorporate indigenous perspectives on biodiversity and consent to minimize risks beyond simple ABS mechanisms. To systematically tackle this broad threat and the policy paralysis it reveals, this chapter introduces the Policy-to-Action framework, which provides practical guidance for unlocking the bio-economic potential of genome editing. This framework structures the following analysis into four interconnected pillars: (1) governance and law; (2) technology transfer; (3) financing and incentives; and (4) continental integration (**Figure 2**).

3. Governance and law: Building trust, clarity, and confidence

Governance and law provide the essential legal certainty needed to navigate the complexities of genome editing (GEd) and attract investment. This pillar has two key roles: first, to achieve regional harmonization that breaks down IPR-based non-tariff barriers; and second, to create strong ethical and equity frameworks that manage monopolistic risks and guarantee public access to important technologies. Addressing these issues involves prioritizing the integration of access and benefit-sharing (ABS) mechanisms and clarifying regulatory alignment.

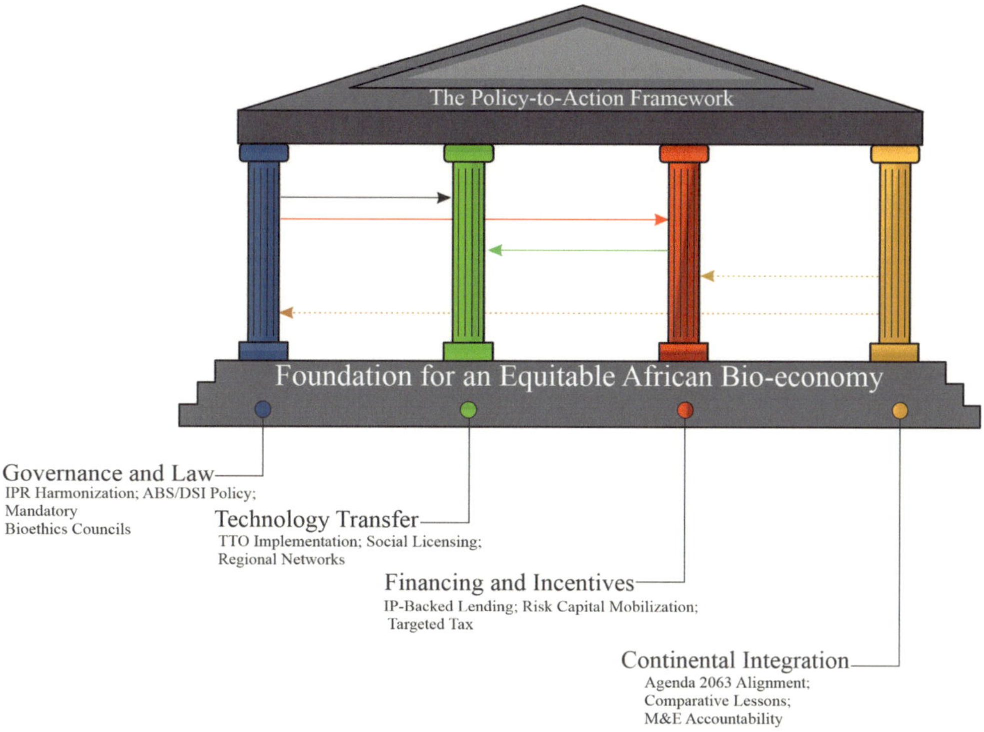

Figure 2.
The four pillars of the policy – to-action framework (p–t–af) and their causal interdependencies. The policy-to-action framework is designed as a dynamic, systemic solution where the pillars operate in a sequence of dependencies. The core execution engine flows logically, beginning with pillar 1 (governance and law) and then proceeding to pillar 3 (financing and incentives), demonstrating that legal certainty and strong governance are preconditions for attracting and mobilizing capital. This flow continues with pillar 3 (financing) and pillar 2 (technology transfer), as capital and resources provide the necessary fuel to enable TTOs to scale their commercial operations. Providing top-down oversight and strategic direction, the alignment and accountability layer is driven by pillar 4 (continental integration). This layer aligns both the policy agenda (pillar 1) and the financial mechanisms (pillar 3) to ensure their actions ultimately serve the unified regional market objectives of the bio-economy.

3.1 Harmonizing ethical standards: Access, benefit-sharing, and DSI

The most urgent and longstanding challenge is the exploitation of genetic resources, often called biopiracy. Africa's genetic resources have been exploited for a long time without fair compensation. Although the Nagoya Protocol [45] aimed to address this through Access and Benefit-Sharing (ABS) frameworks, enforcement remains weak and highly fragmented across the continent. Genome editing (GEd) adds further complexity with data-driven bioeconomies. Genetic information, known as Digital Sequence Information (DSI), can now be stored, transferred, and patented digitally abroad, bypassing traditional physical borders and existing ABS systems. This loophole presents an immediate and critical challenge for the Policy-to -Action Framework to establish unified policies on DSI and ABS enforcement, specifically to prevent the ongoing exploitation of Africa's genetic resources.

To complement the ethical frameworks necessary for DSI and ABS, Pillar 1 should also focus on the Action Area of IPR Harmonization and TRIPS Flexibilities. The current legal environment, managed by regional bodies such as the African Regional Intellectual Property Organization (ARIPO) and the Organisation Africaine de la Propriété Intellectuelle (OAPI), is inconsistent and lacks clarity for high-value GEd investment. Therefore, the main goal of the Framework is to promote regional harmonization of IP laws that are adaptable enough to align with African realities. This harmonization must actively incorporate the public interest safeguards allowed under the World Trade Organization's (WTO) TRIPS Agreement. African policymakers need to learn from global debates on access to essential medicines during the HIV/AIDS crisis. Just as the high costs of patented medicines limit access to effective treatments, the high costs of genomic patents will similarly restrict access to vital agricultural and health solutions. Evidence indicates that generic competition is the most effective way to achieve long-term and significant price reductions, but patents greatly hinder this competition.

Therefore, the Framework strongly recommends protocols that clearly allow the use of compulsory licensing during public health crises or food security emergencies. Compulsory licensing enables a state to authorize the production or purchase of generic alternatives without the patent owner's approval, providing a vital tool to bypass monopolies during national emergencies or to support public health and food security. Although these flexibilities remain underused and their implementation can be complex, their explicit inclusion in national laws is crucial to demonstrate political will to prioritize the public interest over exclusive private IPR. Additionally, to effectively encourage local innovation and entrepreneurship, national policies should simplify the process for granting utility model patents – a shorter, less strict form of protection best suited for incremental local inventions.

Achieving trust and confidence in GEd technologies requires governance mechanisms that are both ethically proactive and judicially robust. The governance of GEd requires that ethical considerations be structurally embedded rather than treated as an ex-post review. This necessitates the mandatory establishment of Bioethics Councils under national science and health commissions. These councils must be institutionally empowered to vet all GEd projects for community consent and ecological safety, which necessitates a mandatory Ethical Impact Assessment (EIA) before any field or clinical applications, particularly for complex and high-impact technologies such as gene drives. Furthermore, the ethical policy must be explicitly grounded in an appropriate African ethical framework, such as Ubuntu (a concept emphasizing "humanity toward others" and interdependence), to mandate shared benefit and collective responsibility. This institutionalizes the principle that scientific benefits must serve the community before private profit, thereby integrating the equity component directly into the regulatory process.

To address the ongoing regulatory fragmentation that hampers innovation, the Policy-to-Action Framework requires the establishment of National Genome-Editing Coordination Units (NGECUs). These NGECUs should serve as a single decision-making authority responsible for harmonizing policies across key ministries, including Agriculture, Health, Environment, and Trade. This coordinated approach is crucial to providing a consistent, predictable regulatory environment for innovators, thereby minimizing conflicts among rules. Additionally, ensuring legal certainty involves addressing the judicial bottleneck. Investor confidence depends on the assurance of quick and informed legal recourse; therefore, governments must commit to strengthening judicial capacity to manage complex biotechnology disputes by establishing specialized biotechnology tribunals or chambers within national courts. This effort ensures that the resolution of complex intellectual property (IP) and regulatory issues is prompt and well-informed, reducing uncertainty and delays that can deter private capital investment.

Finally, effectively managing traditional and digital genetic wealth requires immediate steps to close gaps created by GEd. The Framework's third mandate emphasizes the protection of TK and the regulation of DSI [46]. Legal systems must be updated to recognize, document, and safeguard TK linked to genetic resources, ensuring TK holders are compensated when their knowledge is used in GEd applications. Importantly, to enforce the Nagoya Protocol and oversee the flow of DSI, the Framework proposes establishing a Regional ABS Registry. This centralized registry should integrate with the compliance systems of WIPO and the Nagoya Protocol to monitor the use of African genetic resources, including DSI transferred abroad, and to ensure that fair compensation reaches the source community or country. This systemic step is vital to stop ongoing biopiracy via digital means.

3.2 Technology transfer: Building institutional capacity for equitable innovation

The successful transition of genome-editing innovations from the laboratory to the commercial market depends on a dedicated institutional and policy framework that can translate scientific achievements into lasting economic and social benefits. This process is essential for promoting equity in African nations by addressing key developmental challenges, such as funding gaps, limited local manufacturing capacity, and restricted access to global distribution networks. By building local capacity, this element of the Policy-to-Action Framework ensures that the benefits of GEd technologies are accessible to a wide range of organizations, not just foreign ones. The primary responsibility for this critical translation mechanism lies with Technology Transfer Offices (TTOs), whose success relies on supportive policy reforms.

To fulfill this core function, the policy-to-action framework requires strengthening and the strategic deployment of TTOs as the primary institutional facilitators of equitable innovation. Currently, many African TTOs mainly serve as administrative or patent-processing units, often fragmented across research institutions. They need to be rethought and better supported to go beyond simple IP management and actively engage in technology marketing and entrepreneurship. This change emphasizes two key priorities. First, Mandatory Social Licensing should be introduced, requiring TTOs to include social impact criteria in all licensing agreements for genome-editing (GEd) technologies arising from publicly funded research. It ensures critical public health or agricultural solutions are licensed at preferential or royalty-free rates for humanitarian or non-commercial local research, directly addressing ethical concerns related to monopolistic control. Second, developing professional and commercial skills is essential. The ongoing skills gap in TTOs – particularly in areas such as IP valuation, technology roadmapping, and commercial negotiations – must be addressed. The framework proposes targeted capacity-building initiatives, including international mentorship and a regional TTO certification program, to professionalize the sector and enhance commercial expertise.

To bring genome editing innovations to market, the Policy-to-Action Framework advocates for the strategic use of two main commercialization models: licensing and partnership models, and the creation of spin-offs and start-ups. For the framework to succeed, these strategies must include mechanisms for risk mitigation. In the Licensing and Partnership Models area, agreements allow innovators to grant rights to third-party companies in exchange for compensation, which is suitable for innovators with limited resources. However, the framework requires TTOs to actively address key challenges. A crucial policy action is to equalize bargaining power by supporting TTOs with specialized legal expertise to prevent exploitative agreements. Additionally, TTOs must mitigate risks from weak IPR enforcement. Licensing strategies should account for the inadequate IP protection mechanisms across the continent (as discussed in Section 1) by using strict contractual clauses and solid monitoring. Furthermore, TTOs need to guard against overreliance on partners to ensure that local innovators retain leverage and can explore alternative commercialization paths, maintaining strategic control over technology deployment and ensuring long-term local benefits.

For spin-offs and start-ups, the formation of new ventures serves as a crucial mechanism for leveraging academic research to develop innovative applications. By fostering institutional entrepreneurship, these ventures drive biotechnology commercialization, strategically encouraging risk-taking and creating highly skilled local jobs. However, realizing the full potential of this model requires the framework to take policy action to dismantle systemic barriers. These challenges include overcoming limited access to critical seed funding, rectifying inadequate incubation infrastructure, and addressing persistent regulatory hurdles. This targeted policy intervention is necessary to ensure that local discoveries successfully transition into scalable, sustainable bio-economic enterprises.

Successful commercialization depends on a shared ecosystem of knowledge, infrastructure, and human capital. The Policy-to-Action Framework, therefore, requires establishing collaborative and developmental initiatives. Building on localized successes and leveraging existing regional efforts – such as the African Biosafety Network of Expertise (ABNE) and the Southern African Network for Biosciences (SANBio) – the Framework aims to model approaches that have effectively integrated equity. A key example is Kenya's BioInnovate Africa platform (hosted at ILRI), which successfully brings together scientists, lawyers, and entrepreneurs from East Africa to co-develop biotechnology products under pre-arranged equitable IP agreements. This effective regional collaboration provides the basis for the Framework's proposal: establishing an African Genome Editing Network. This vital network would strategically pool essential IP, share costly equipment, and coordinate research initiatives focused on regional priorities. Importantly, to boost local capacity and reduce reliance on foreign expertise for managing this ecosystem, the Framework immediately calls for dedicated training programs for the human capital needed at the intersection of science and commerce. This specialized talent development should focus on IPR-savvy lawyers, TTO managers, and regulatory scientists. By consolidating and expanding these programs, Africa can develop the expertise needed to manage complex patent portfolios and effectively negotiate equitable, high-value commercial agreements that will drive the bio-economy.

Achieving this level of TTO efficiency and commercial focus requires clarity in institutional policy. To this end, the Framework advocates adopting a principle inspired by international best practices, such as the U.S. Bayh-Dole Act [47], that grants research institutions a clear title to intellectual property (IP) generated from government-funded research. This decisive policy measure provides a clear, uniform incentive for TTOs to manage and commercialize their portfolios aggressively, reversing the trend in which valuable IP remains "stuck in the laboratory" or is ceded prematurely. Without this institutional policy clarity, even scientifically excellent research will likely fail to achieve its full potential for economic utilization and public benefit.

3.3 Financing, incentives, and sustainability: Mobilizing capital for the bio-economy

This subsection transitions from establishing legal and institutional structures (Pillars 1 and 2) to defining the practical financial incentives and mechanisms required to secure and sustain a vibrant African biotech ecosystem (**Figure 2**). The core challenge lies in addressing the persistent lack of risk capital for high-tech, multi-year GEd ventures. The Policy-to-Action Framework mandates a three-pronged approach to mobilize both public and private capital.

3.3.1 Mobilizing risk capital via strategic funds

The initial phase of genome-editing projects often faces challenges due to the high risks and long timelines involved in biotech R&D, which make obtaining commercial bank loans difficult. To address this, the Framework proposes establishing National Innovation Funds with a dedicated high-risk biotech component. These funds, possibly supported by regional development banks like the African Development Bank (AfDB), would provide catalytic seed funding and early-stage venture capital. The aim is to cover the initial financial risks of local spin-offs and academic ventures, thereby bridging the funding gap until the technologies are sufficiently de-risked to attract traditional private equity.

3.3.2 De-risking investment through IP assets

To boost private-sector involvement and lessen dependence on external investment, policies should be put in place to reduce risks for local commercial loans by utilizing the value of intellectual property. The framework specifically suggests creating IP-Backed Lending Guarantee Schemes. This strategy is essential for transforming intangible assets into financial collateral. It involves establishing standardized procedures, in collaboration with national patent offices, to assess IP assets (such as patents, registered know-how, and proprietary data). Once evaluated, these assets can be used as collateral for loans. By offering government or development bank guarantees for these IP assets, the credit risk for local banks is greatly lowered, encouraging private-sector investment and allowing local innovators to access substantial capital without giving up equity or control.

3.3.3 Fiscal measures and R&D incentives

Supporting the bio-economy requires consistent, long-term fiscal signals that encourage and facilitate commercialization and local production. The Framework, therefore, recommends specific, targeted tax incentives designed to motivate both domestic and foreign companies to invest in local R&D and manufacturing. These essential incentives should include offering R&D Tax Credits, which provide substantial tax deductions or credits for expenses related to GEd research conducted within the country. This must be complemented by providing Tax Holidays, offering temporary tax relief for profits generated from the local manufacturing and commercialization of vital biotechnologies, such as drought-resistant seeds or diagnostic tools. Collectively, these fiscal measures create an environment that significantly lowers operational costs for high-tech firms, anchoring R&D and production activities within Africa, thereby ensuring sustainability and maximizing the local economic benefits derived from genomic innovation.

3.4 Continental integration and strategic alignment

The successful implementation of the Policy-to-Action Framework depends on its recommendations having both continental legitimacy and global credibility. This final pillar ensures the Framework aligns strategically with the African Union's long-term development goals, learns valuable lessons from successful international examples, and commits to clear accountability. This integration is vital to gain the high-level political support and investment needed for effective implementation.

3.4.1 Ensuring continental legitimacy

Implementing the Framework's mandates requires backing from top continental leaders. Therefore, the Framework should be seen not just as a technical document but as a tool directly supporting important socio-economic goals. To gain this support, the main strategy involves clearly linking the Framework's key recommendations – such as the Regional ABS Registry, IP-Backed Lending Schemes, and NGECUs – to the African Union's main documents. Specifically, these policies must be shown to be relevant by connecting them to high-level goals within Agenda 2063 (especially Aspiration 1: "A Prosperous Africa") and the Science, Technology, and Innovation Strategy for Africa (STISA-2024). This clear connection is crucial for justifying resource use, building political credibility, and maintaining consistent support across the continent for the Framework's rollout.

3.4.2 Comparative lessons and global blueprints

To improve the chances of successful policy design, the Framework's recommendations must be based on successful global blueprints, especially those from developing countries facing similar challenges in balancing access and innovation. The Framework supports expanding the comparative analysis to include models that effectively implement flexibilities relevant to the African context. India's Patent Act, which effectively limits patent extensions (evergreening) and ensures public access to essential technologies, provides an important example for balancing IPR protection with public health safeguards. At the same time, Brazil's comprehensive biodiversity laws offer relevant regulatory frameworks and legal experience for developing nations aiming to manage genetic resources and enforce benefit-sharing agreements effectively. By referencing these specific, proven non-Western models, the Framework emphasizes the feasibility of a system that prioritizes public welfare without discouraging investment.

3.4.3 Monitoring, evaluation, and accountability

The long-term sustainability and effectiveness of the Policy-to-Action Framework depend on a robust system of accountability. This involves clearly defined success metrics that go beyond merely adopting policies to measure tangible bio-economic results. Governments and regional organizations implementing the Framework must commit to a set of quantitative indicators to assess its success. These key performance indicators (KPIs) should address three main areas: Institutional Capacity, evaluated by the number of fully operational, adequately funded TTOs; Innovation Output, measured by the increase in high-quality biotech patents filed locally and the number of successful spin-offs from TTOs; and Commercial Impact, gauged by the growth in local innovation commercialization rates and the volume of successful IP-backed loans. By following these measurable M&E indicators, the Framework provides a clear path for continuous improvement and ensures the political accountability necessary to build an equitable African bio-economy.

4. Conclusion

Africa's adoption of genome editing offers a key chance for significant growth in its food, health, and industrial sectors, but only if strong legal, institutional, financial, and ethical systems are put in place. This can be achieved by using TRIPS flexibilities, such as compulsory licensing and social licensing, while carefully designing IPR mechanisms that prioritize public needs over private rights. Additionally, ethical governance must be institutionalized and based on African principles to prevent biopiracy, protect community consent, manage new threats like Digital Sequence Information (DSI), and address disparities in sequencing costs to improve access. Recent developments in 2025, including CGIAR's genome editing workshops in Nairobi and UCT's groundbreaking gene-editing trial for hereditary angioedema, show increasing momentum but also highlight the need for harmonized policies to avoid monopolies in a global market expected to grow beyond $13 billion. Ultimately, expanding local innovation through continental cooperation and IP-backed lending is essential for reaching critical mass and remaining competitive worldwide, aligning with the African Union's Agenda 2063 vision. Policymakers must urgently implement the Policy-to-Action Framework to ensure Africa actively shapes its bioeconomic future and sparks worldwide discussions on fair biotech progress.

Author details

Muyiwa S. Adegbaju[1*], Olalekan Akinbo[2], Rachel Shibalira[3], Olufunke B. Shittu[4], Ruth Asheazi Manzah[5], Sunday Igu Rock Akile[2], Charles Oluwaseun Adetunji[6] and Bolaji N. Thomas[1]

1 Rochester Institute of Technology, Rochester, NY, United States of America

2 Directorate of Agriculture, Food Security and Environmental Sustainability, African Union Development Agency – NEPAD (AUDA-NEPAD), Midrand, South Africa

3 Independent Policy and Legal Consultant specializing in Agri-Food Systems, Environmental Governance, and Public Sector Policy, Nairobi, Kenya

4 Federal University of Agriculture, Abeokuta, Nigeria

5 National Biosafety Management Agency, Abuja, Nigeria

6 Department of Microbiology, Edo State University, Auchi, Nigeria

*Address all correspondence to: msabms@rit.edu

References

[1] AFSLD. Communiqué: Africa Food Security Leadership Dialogue. 2019. https://documents1.worldbank.org/curated/en/345881572619261427/pdf/Communique-Africa-Food-Security-Leadership-Dialogue.pdf [Accessed: 2026-January-15]

[2] Gatune J, Ozor N, Oriama R. The futures of bioeconomy in Eastern Africa. Journal of Futures Studies. 2021;**25**(3):1–14

[3] Gatune J, et al. The futures of bioeconomy in Eastern Africa. Journal of Futures Studies. 2021;**25**:1–14

[4] Malabo Montpellier Panel. Nourished: How Africa can build a future free from hunger and malnutrition. 2022. Available from: https://www.mamopanel.org/media/uploads/files/Nourished_How_Africa_Can_Build_a_Future_Free_from_Hunger_and_Malnutrition.pdf [Accessed: 2026-January-15]

[5] Rahman K, Khan SU, Fahad S, Chang MX, Abbas A, Khan WU, et al. Nano-biotechnology: A new approach to treat and prevent malaria. International Journal of Nanomedicine. 2019;**14**:1401–1410

[6] Wang YC, Lee YT, Yang T, Sun JR, Shen CF, Cheng CM. Current diagnostic tools for coronaviruses–From laboratory diagnosis to POC diagnosis for COVID-19. Bioengineering & Translational Medicine. 2020;**5**(2):e10177. DOI: 10.1002/btm2.10177

[7] Behera BC, Mishra RR, Thatoi H. Recent biotechnological tools for diagnosis of Corona virus disease: A review. Biotechnology Progress. 2021;**37**(1):e3078

[8] Fitri LE, Widaningrum T, Endharti AT, Prabowo MH, Winaris N, Nugraha RYB. Malaria diagnostic update: From conventional to advanced method. Journal of Clinical Laboratory Analysis. 2022;**36**(3):e24314

[9] MacLean E, Kohli M, Weber SF, Suresh A, Schumacher SG, Denkinger CM, Pai M. Advances in molecular diagnosis of Tuberculosis. Journal of Clinical Microbiology. 2020;**58**(3):1–13

[10] Kaufmann She. Vaccine development against tuberculosis before and after COVID-19. Frontiers in Immunology. 2023;**14**:1273938

[11] Nair A, Greeny A, Nandan A, et al. Advanced drug delivery and therapeutic strategies for tuberculosis treatment. Journal of Nanobiotechnology 2023;**21**:414

[12] Adegbaju MS, Ajose T, Adegbaju IE, Omosebi T, Ajenifujah-Solebo SO, Falana OY, et al. Genetic engineering and genome editing technologies as catalyst for Africa's food security: The case of plant biotechnology in Nigeria. Frontiers in Genome Editing. 2024;**6**:1398813

[13] Nature Editorial. Licensing CRISPR patents for free to share gene editing globally. Nature. 2021;**597**(7875):152

[14] Global Innovation Index 2025-WIPO

[15] Entine J, Felipe M, Groenewald JH, Kershen DL, Lema M, McHughen A,

et al. Regulatory approaches for genome edited agricultural plants in select countries and jurisdictions around the world. Transgenic Research 2021;**30**:551–584

[16] AAGEPF. AUDA-NEPAD APET genome editing policy framework. AAGEPF. 2022. Available from: https://assets.au-apet.org/knowledge-products/genome-editing/PolicyFrameworkforApplicationsofGenomeEditing.pdf [Accessed: 2025-October-16]

[17] Akinbo O, Nkhabindze B, Amedu J, Ebegba R, Asagbra A, Ratemo BO, et al. Africa and zero hunger agenda: Genome editing policy landscape, challenges and opportunities.Frontiers in Bioengineering and Biotechnology. 2025;**13**:1526851

[18] WIPO. Innovation Gender Gap: Gender Stories. World Intellectual Property Organization. 2024. Available from: https://www.wipo.int/en/web/economics/innovation-gender-gap/insights [Accessed: 2026-January-15]

[19] Rock JS, Schnurr MA, Kingiri A, Ely A, Glover D, Stone GD, Fischer K. The knowledge politics of genome editing in Africa. Elementa: Science of the Anthropocene. 2023;**11**(1):00143

[20] Nwanna CE. Intellectual property rights in Nigeria: Issues and challenges. American Journal of Law and Legal Studies. 2020;**3**:7. DOI: 10.28933/ajlls-2019-12-2505

[21] Kebede D, Zielinski C, Mbondji PE, Sanou I, Kouvividila W, Lusamba-Dikassa P-S. Research output of health research institutions and its use in 42 sub-Saharan African countries: Results of a questionnaire-based survey. Journal of the Royal Society of Medicine. 2014;**107**(1_suppl):105–114

[22] International IP Index 2025. https://www.uschamber.com/assets/documents/GIPC_IPIndex2025_Combined_final.pdf [Accessed: 2026-January-15]

[23] Beer JD, Baarbé J, Ncube CB. Evolution of Africa's intellectual property treaty ratification landscape. The African Journal of Information and Communication. 2018;**22**:53–82

[24] Mhaka TW. The impact of intellectual property rights on regional integration in Africa: Opportunities and challenges. International Journal of Academic Multidisciplinary Research. 2024;**8**(7):436–439

[25] Glatzel K, Virchow D, Nakitto AMS, Niyonsenga S, Babu S, Srivastava N, et al. Bioeconomy pathways: Experience from Africa, Asia, and Latin America. In the 2024 Annual Trends and Outlook Report: Advancing the Climate and Bioeconomy Agenda in Africa for Resilient and Sustainable Agrifood Systems. 2024

[26] Chidede T. The role of intellectual property rights' protection in advancing development in South Africa. Law, Democracy & Development. 2022;**26**:168–189

[27] Hanauer LH, Ramaphala IQ. IP regulation in South Africa evolution, current status, and comparison with SADC Region Countries. Revista La Propiedad Inmaterial. 2020;**30**:548

[28] Akinbo O, Shittu OB, Adetunji CO, Adegbaju MS, Gbodope SI, Ajenifujah-Solebo SO, et al. Advancing policy and science: A game changer in the adoption of genome editing technology in the

continent of Africa. ACS Agricultural Science & Technology. 2025;**5**(7):1–25

[29] Zimu-Biyela AN. What is the role of libraries in disseminating knowledge about South African intellectual property laws in rural communities? South African Journal of Libraries and Information Science. 2021;**87**(2):21–29

[30] Sackey NA, Boahen EA, Sackey E. The role of intellectual property as a tool for business development in the informal economy of Ghana. The Journal of World Intellectual Property. 2024;**28**(2):1–22

[31] Ampaire E, Nyabwanga R, Otieno A. Intellectual property rights and innovation: Evidence from ICT firms in Kenya. Journal of Innovation and Entrepreneurship. 2020;**9**:1–17

[32] Nweke OC, Kadogbe C.The role of intellectual property in Ghana's growing startup ecosystem: How IP protection fosters innovation and entrepreneurship. Law and Humanities Quarterly Reviews. 2025;**4**(3):57–69

[33] Afolayan OT. Intellectual property rights protection in Nigeria: Issues and perspectives. Information Impact: Journal of Information and Knowledge Management. 2022;**13**(1):1–9. DOI: 10.4314/iijikm.v13i1.1

[34] Monsanto Technology LLC v Cargill International SA. (2007) EWHC 2257(Pat)

[35] Keith Kirsten's (Pty) Limited v Weltevrede Nursery (Pty) Limited and Another. (2002). [2002] ZAWCHC 18

[36] Abuduxike G, Aljunid SM. Development of health biotechnology in developing countries: Can private-sector players be the prime movers? Biotechnology Advances. 2012;**30**(6):1589–1601

[37] Özkan N. Toplum sağlığının hizmetinde bilim ve teknoloji: Küba'da biyoteknoloji deneyimi [Science and technology at the service of community health: Experience of biotechnology in Cuba]. Madde, Diyalektik Ve Toplum. 2019;**2**(1):8–18

[38] Kameri-Mbote P, et al. Public funding for agriculture in Africa: Challenges and reforms. African Journal of International Law. 2001;**9**:45–67

[39] Burk DL. Biotechnology and patent law: Fitting innovation to procrustean bed. Rutgers Computer and Technology Law Journal. 1991;**17**(1):1–85

[40] Safrin S. Hyperownership in a time of biotechnological promise: The international conflict to control the building blocks of life. The American Journal of International Law. 2004;**98**(4):669

[41] Clark V. Pitfalls in drafting royalty provisions in patent licences. Bio-Science Law Review. 2003;**6**(2). Available from: https://www.lawtext.com/pdfs/sampleIssues/Bio_Science_6.2.pdf

[42] Open AIR. Working around the Gender Gap in Intellectual Property Regimes. Open African Innovation Research; 2020. Available from: https://openair.africa/wp-content/uploads/2020/07/WP-23-Working-around-the-Gender-Gap-in-Intellectual-Property-Regimes.pdf

[43] Burgess MM, Mumford KL, Chowns AR. Empowering indigenous knowledge in deliberations on gene

editing in the wild. The Hastings Center Report. 2021;**51**(S2):S74–S84

[44] Taitingfong R, Ullah A. Empowering Indigenous Knowledge in Deliberations on Gene Editing in the Wild. Hastings Center Report. 2021;**51** (2):S74–S84. DOI: 10.1002/hast.1323

[45] Beato MS, Veneroso V. The Nagoya Protocol on access and benefit sharing: The neglected issue of animal health. Frontiers in Microbiology. 2023;**14**:1124120. DOI: 10.3389/fmicb.2023.1124120

[46] Taitingfong R, Ullah A. Empowering indigenous knowledge in deliberations on gene editing in the wild," in gene editing in the wild: Shaping decisions through broad public deliberation Michael K G.ed., et al, special report. Hastings Center Report. 2021;**51**(S2):S74–S84. DOI: 10.1002/hast.1323

[47] Bayh-Dole Act, 35 U.S.C. §§. 1980:200–212